EPA/100/B-15/001

U.S. Environmental Protection Agency

Peer Review Handbook

4th Edition

October 2015

Prepared for the U.S. Environmental Protection Agency
under the direction of the EPA Peer Review Advisory Group

Science and Technology Policy Council
U.S. Environmental Protection Agency
Washington, D.C. 20460

DISCLAIMER

This 4th edition of the *Peer Review Handbook* was developed by the U.S. Environmental Protection Agency (hereafter EPA or the Agency) to provide guidance to EPA staff and managers who are planning and conducting peer reviews. It is intended to improve the internal management of EPA peer review by providing recommended procedures and approaches for EPA staff and managers. This 4th edition is a guidance manual and not a rule or regulation. Some topics in the Handbook refer to laws or EPA policies. In such cases, this Handbook provides recommendations for how those provisions can be implemented. The *Peer Review Handbook* does not replace existing laws or regulations, does not change or substitute for any legal requirement, and is not legally enforceable. This 4th edition does not create or confer legal rights or impose any legally binding requirements on EPA or any party. The use of non-mandatory language such as "may," "can" or "should" in this *Peer Review Handbook* does not connote a requirement but does indicate EPA's strongly preferred approach to ensure the quality of peer reviews conducted or initiated by EPA. Mention of trade names or commercial products does not constitute endorsement or recommendation for use.

PEER REVIEW HANDBOOK WRITING GROUP

Mary E. Clark – Writing Group Chair   OAR

Jane C. Caldwell   ORD
Daniel Fort   OGC
Cheryl A. Hawkins   OSWER
Jeffrey Herrema   OGC
Vincia Holloman   OEI
Virginia Houk – PRAG Chair   ORD
Cheryl Itkin   ORD
Jacques Kapuscinski   ORD
Eric Koglin   ORD
Linda Mauel   Region 2
Anand Mudambi   OSA
Marian Olsen   Region 2
Stephanie Sanzone   SABSO
Tracy Sheppard   OGC

Peer Review Advisory Group (PRAG): Link to the list of members (with their office/region affiliation): http://intranet.ord.epa.gov/about/organization/osa/peer-review-advisory-group.

CONTENTS

LIST OF FIGURES

LIST OF EXHIBITS

LIST OF TABLES

FOREWORD

Science is the foundation that supports all of our work at EPA. The quality and integrity of the science that underlies our regulations are vital to the credibility of EPA's decisions and, ultimately, the Agency's effectiveness in pursuing its mission to protect human health and the environment. One important element in ensuring that decisions are based on sound and defensible science is to have an open and transparent peer review process.

EPA has a long-standing history of peer review. The Agency has been a leader across the federal government in developing guidance and support for the peer review process. Even before issuing its Agency-wide Peer Review Policy in 1993, EPA was committed to peer review of its scientific and technical products. Over the years, EPA has repeatedly reaffirmed and updated both its Peer Review Policy and the processes for implementing peer review to ensure that EPA decisions rest on credible science and data.

The Agency's *Peer Review Handbook* was first released in 1998 and has been updated several times since. Each update has emphasized greater transparency and accountability for peer review. The last edition of the Handbook (2006) incorporated the provisions of the Office of Management and Budget's (OMB) *Final Information Quality Bulletin for Peer Review*. An EPA Addendum to the Handbook in 2009 provided guidance on preventing ethics concerns related to the appearance of a loss of impartiality for peer reviewers.

This newly revised 4th edition of the *Peer Review Handbook*, commissioned by the EPA Science and Technology Policy Council (STPC), supersedes all previous editions. Although the basic peer review procedures in the 2006 *Peer Review Handbook* remain current and our overall approach to peer review is not changing, this revision enhances and reinforces the practice of peer review at the Agency.

This *Peer Review Handbook* should be used as guidance by EPA staff and managers to ensure that the Agency's Peer Review Policy is implemented effectively and that the integrity of our peer review activities can be demonstrated transparently to the American public.

Thomas A. Burke, PhD, MPH
EPA Science Advisor

PREFACE

The first edition of the EPA *Peer Review Handbook* was issued in 1998 and was intended to serve as a single, centralized source of implementation guidance on peer review for EPA staff and managers. Subsequent revisions of the Handbook have added necessary clarifications, incorporated insights and experiences gained through its use, and integrated changes to reflect updated government-wide guidance or policy related to peer review. These revisions have increased the transparency and accountability of peer review and helped ensure that Agency decisions are based on sound and defensible science.

For the 4th edition, the EPA's STPC determined that revisions were needed to incorporate several recent EPA policy and process changes related to peer review. Although the 4th edition draws heavily from the 3rd edition, it has been reorganized to emphasize the elements and tools needed to implement a systematic peer review. It retains, however, the "question and answer" format throughout. New flowcharts and checklists have been added, and several substantial updates are included, such as the additional guidance on appearance of a loss of impartiality in external peer reviews, new information on organizational changes and oversight responsibilities, and changes related to the issuance of recent policies and procedures associated with the EPA's Information Quality Guidelines (IQG). The 4th edition also describes process changes for contractor-managed panel peer reviews of scientific and technical documents designated as Influential Scientific Information (ISI), including Highly Influential Scientific Assessments (HISAs), which are a subset of ISI. The process is intended to reduce the potential for organizational or personal conflict-of-interest (COI) concerns. Early public participation in the nomination and selection of peer reviewers and increased internal oversight are features of the process.

As in previous editions of the Handbook, not every peer review scenario can be anticipated or discussed. Through the use of examples, tools (e.g., flow diagrams, checklists) and process descriptions, however, this 4th edition illustrates practices from across the Agency that demonstrate effective implementation of peer review policy. The use of the recommended procedures and approaches in this Handbook should reinforce the open, transparent and objective peer review of Agency products.

ABBREVIATIONS AND ACRONYMS

AA   Assistant Administrator
ADP   Action Development Process
CASAC   Clean Air Scientific Advisory Committee
CBI   Confidential Business Information
CO   Contract(ing) Officer
COI   Conflict of Interest
COR   Contracting Officer's Representative
DA   Deputy Administrator
DAEO   Designated Agency Ethics Official
DEO   Deputy Ethics Official
DFO   Designated Federal Officer
DM   Decision Maker
DQA   Director of Quality Assurance
EIS   Environmental Impact Statement
EPA   U.S. Environmental Protection Agency
EPAAG   EPA Acquisition Guide
FAC   Federal Advisory Committee
FACA   Federal Advisory Committee Act
FAR   Federal Acquisition Regulations
FOIA   Freedom of Information Act
FTE   Full-Time Equivalent
GSAPR   Gratuitous Services Agreement for Peer Review
HISA   Highly Influential Scientific Assessment
IGA   Inherently Governmental Activity
IQG   Information Quality Guidelines
IRIS   Integrated Risk Information System
ISI   Influential Scientific Information
NAS   National Academy of Sciences
NCEA   National Center for Environmental Assessment
NEPA   National Environmental Policy Act
NRC   National Research Council
NTTAA   National Technology Transfer and Advancement Act of 1995
OGC   Office of General Counsel
OGE   U.S. Office of Government Ethics
OMB   Office of Management and Budget
ORC   Office of Regional Counsel
ORD   Office of Research and Development
OSA   Office of the Science Advisor
PI   Principal Investigator
PL   Project Leader
PM   Project Manager
PRAG   Peer Review Advisory Group
PRC   Peer Review Coordinator
PRL   Peer Review Leader
QA   Quality Assurance
QAM   Quality Assurance Manager
RA   Regional Administrator

RGE	Regular Government Employee
ROD	Record of Decision
SAB	Science Advisory Board
SAP	Scientific Advisory Panel
SGE	Special Government Employee
SI	Science Inventory
SOW	Statement of Work
STPC	Science and Technology Policy Council

ROADMAP TO PEER REVIEW AT EPA

ROADMAP TO PEER REVIEW AT EPA

R.1. Overview

The goal of this roadmap is to assist the user in understanding how to apply the material in the Handbook and determining where important decisions should be made and documented. Figure 1 summarizes the Agency's overall peer review process, whereas Figures 2 and 3 provide additional details of the key steps, decisions and milestones. This roadmap is not meant to be a stand-alone document but is to be used as a quick reference to users already familiar with the systematic process of planning, conducting and completing peer reviews. Roadmap users will find flowcharts summarizing major decision points in the process and times where documentation is needed, with references to specific sections in the Handbook containing more detailed information. Although the roadmap assumes familiarity with general Agency terminology, Section 1.2 of the Handbook discusses key terms associated with this guidance.

This roadmap also includes example tools for (1) documenting peer review decisions; (2) developing regulatory action; and (3) planning, conducting and completing the peer review. Because these tools vary depending on both the intended use of the work product and the decisions to be made, more than one tool generally is needed.

R.2. Relationship between the Roadmap and Chapters 1 Through 7

The roadmap figures show the peer review process from start to finish. The Handbook Chapters 1 through 7 have been organized to describe essential elements and concepts (the "what") needed for successful implementation of the peer review process. General concepts included are:

- providing terms and context (see Chapter 1);

- identifying relevant peer review roles, responsibilities and resource considerations of Agency personnel and organizations (see Chapter 2);

- categorizing work products (see Chapter 3);

- determining the appropriate peer review approach (see Chapter 4);

- selecting reviewers and considering associated ethics issues such as potential conflicts of interest (COIs) or an appearance of a loss of impartiality (see Chapter 5);

- conducting and completing the review, including developing the peer review charge (see Chapter 6); and

- ensuring transparency during various steps in the peer review process (see Chapter 7).

For some, the process may be described more effectively visually, using diagrams or graphics to make relationships more apparent and provide easy navigation through the entire process. Figures 1 through 3 are the main processes described in this Handbook, provided in graphic form.

Figure 1, the diagram of the peer review process, illustrates the Agency's overall peer review process for scientific or technical (including economic and social science) work products. The Agency process emphasizes early categorization of the work product—preferably at the conceptual stage—into one of three categories: Influential Scientific Information (ISI); Highly Influential Scientific Assessment (HISA), which is a subset of ISI; or other. The ISI and HISA categories have been identified and defined by the Office of Management and Budget (OMB) in its *Final Information Quality Bulletin for Peer Review* (OMB Peer Review Bulletin) (Appendix B). Management approval and documentation of key decisions throughout the peer review process are emphasized. The EPA also demonstrates its commitment to transparency in the peer review process by providing opportunities for public participation.

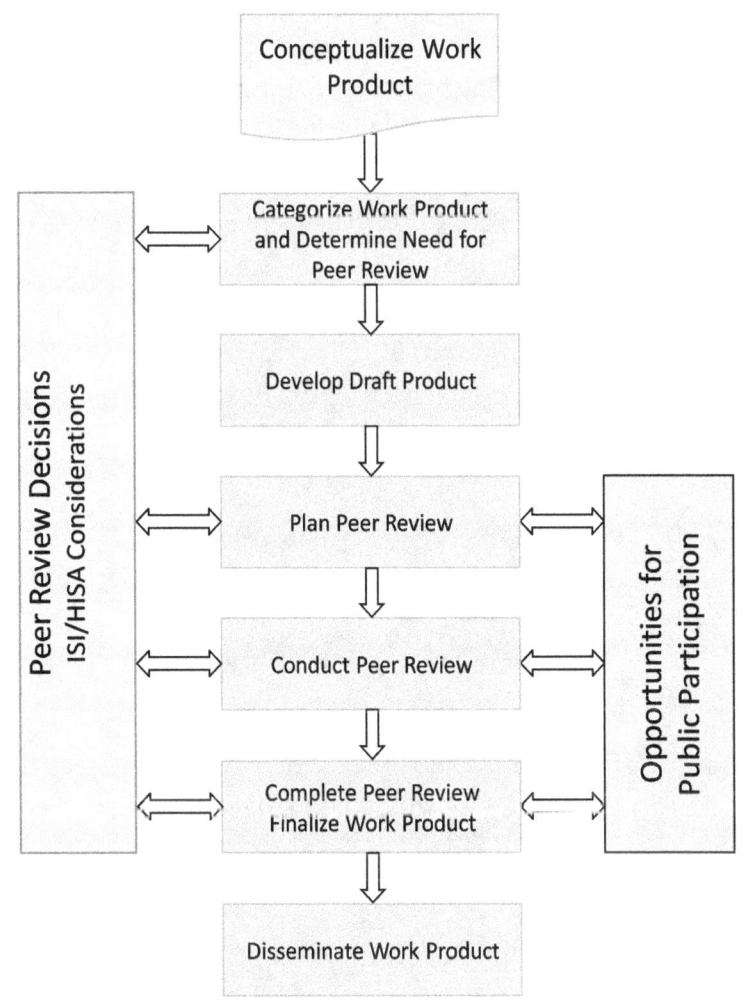

Figure 1. The Peer Review Process

Figure 2, the peer review flowchart for influential work products, illustrates details associated with the general process. Each of the four phases in this flowchart is presented subsequently in Figures 2a through 2d and references to relevant Handbook sections are provided. The figures also include steps at which the Decision Maker (DM) should be involved, and points at which the peer review record, as well as the EPA's searchable database for influential products, the Science Inventory (SI),[1] should be updated. Although updating the SI provides public access to the information about the peer review, the figures indicate various points in the peer review process where the public may also be provided opportunities to comment on materials in the SI.

Figure 3 illustrates the comparable flow for scientific or technical work products not categorized as ISI or a HISA. It includes a specific process for work products that will be submitted to peer-reviewed journals; in that case, work products are subject to management review (following the procedures of the program or regional office) prior to submission to a journal, and authors work with the journal editors/reviewers to resolve any comments. For more information on peer review of work products not categorized as ISI or a HISA, see Sections 3.2.5 and 3.2.6.

---

[1] EPA. 2015. *EPA Science Inventory.* http://cfpub.epa.gov/si/.

It should be noted that the peer review flow charts show the general steps that are followed for the peer review of work products at EPA. The specific steps taken by individual EPA offices will depend on many factors, including the type of work product, timeframe available for peer review and resource considerations. It should be noted that the term "EPA offices" in this Handbook refers to all headquarters, regional and program offices.

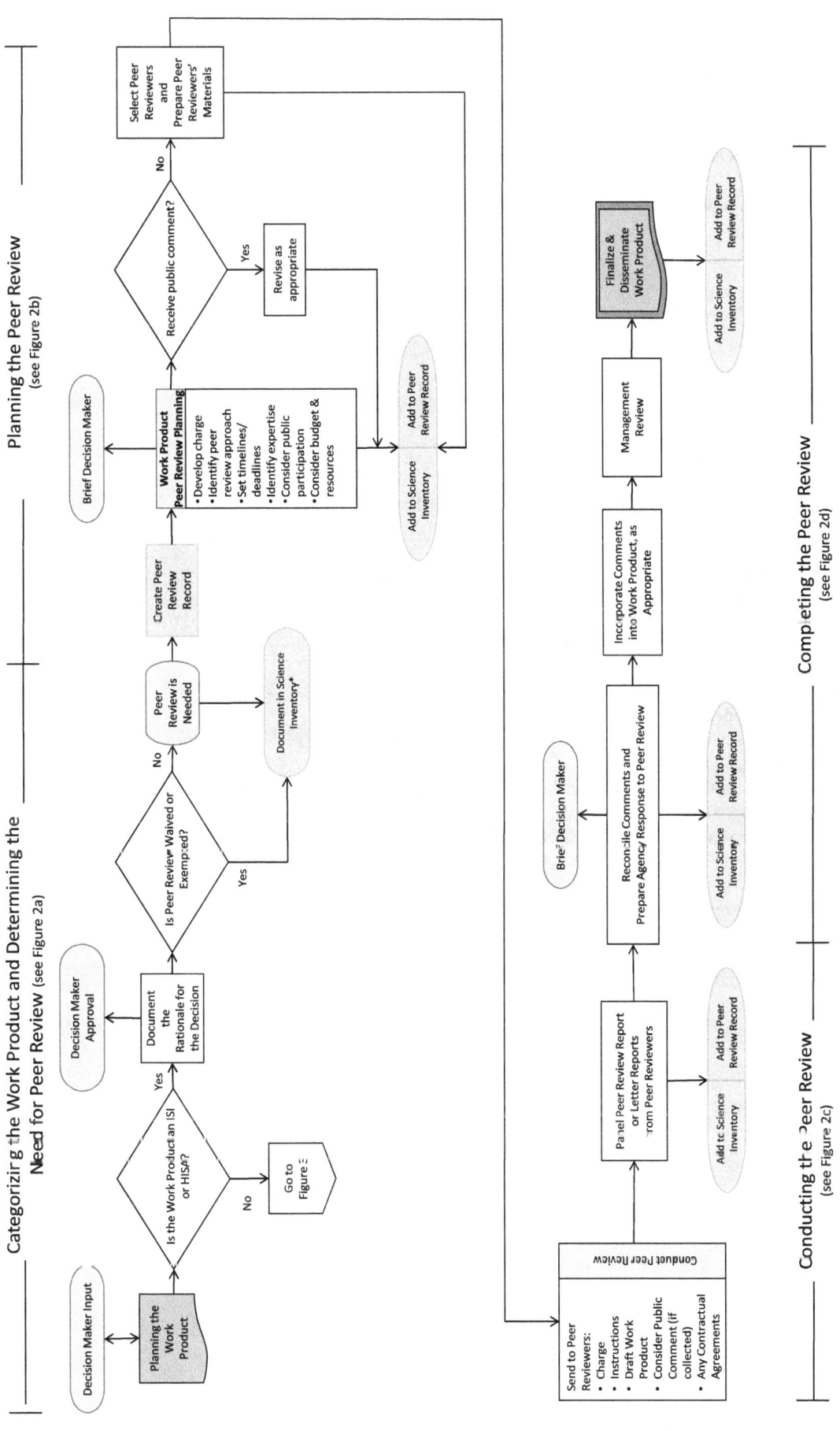

Categorizing the Work Product and Determining the Need for Peer Review (see Figure 2a)

Planning the Peer Review (see Figure 2b)

Conducting the Peer Review (see Figure 2c)

Completing the Peer Review (see Figure 2d)

*Agency's Peer Review Agenda is created from information entered in the Science Inventory

Figure 2. Detailed Peer Review Flowchart for Influential Work Products (Including HISAs)*

* For work products categorized as "other," see Figure 3.

EPA Peer Review Handbook: Roadmap

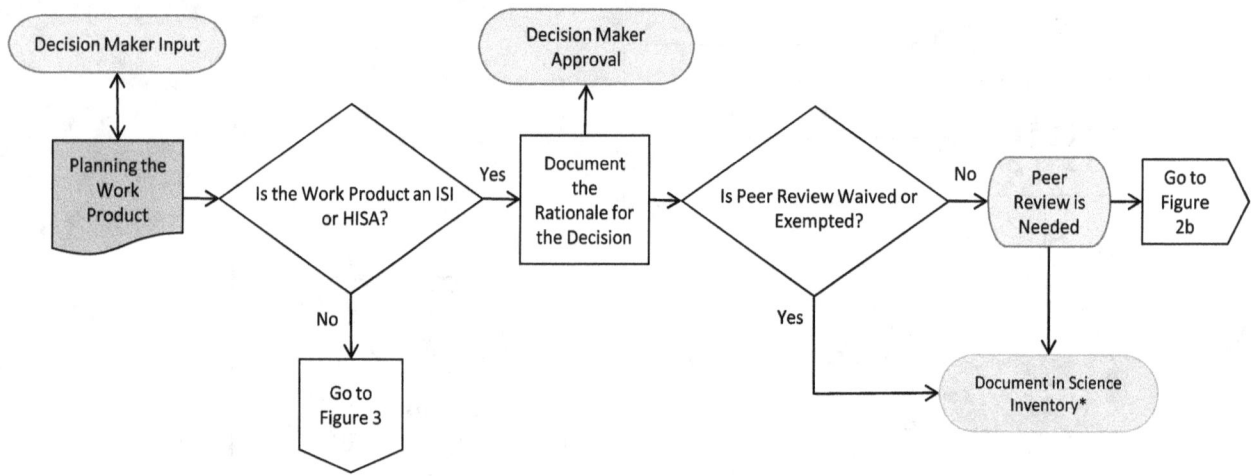

*Agency's Peer Review Agenda is created from information entered in the Science Inventory

---

Figure 2a. Categorizing the Work Product and Determining the Need for Peer Review

1. Determine if the work product:
  - Is a scientific, engineering, economic, social science or statistical document (§ 3.1.1, 3.1.3)
  - Is ISI/HISA (§§ 3.2.1, 3.2.3, 3.2.4)
  - Other work product (see Figure 3)

2. Obtain categorization of work product from the DM:
  - Document decision and rationale for decision
  - Continue with peer review unless determined not to be needed

3. Peer review typically not needed if:
  - ISI/HISA consists only of science previously peer reviewed and the previous peer review is deemed adequate under the Agency's policy (§ 3.3.2)
  - ISI/HISA consists only of principal findings, conclusions and recommendations from National Academy of Sciences (NAS) official reports (Appendix B, Section III.2)
  - Work product meets criteria for exemption (§§ 3.3.1, 3.3.2)
  - Work product receives waiver (§ 3.3.3)
  - Peer review otherwise determined not to be warranted

4. Add document with waiver/exemption to the SI[2]

---

[2] EPA. 2015. *Peer Review Agenda.* http://cfpub.epa.gov/si/si_public_pr_agenda.cfm.

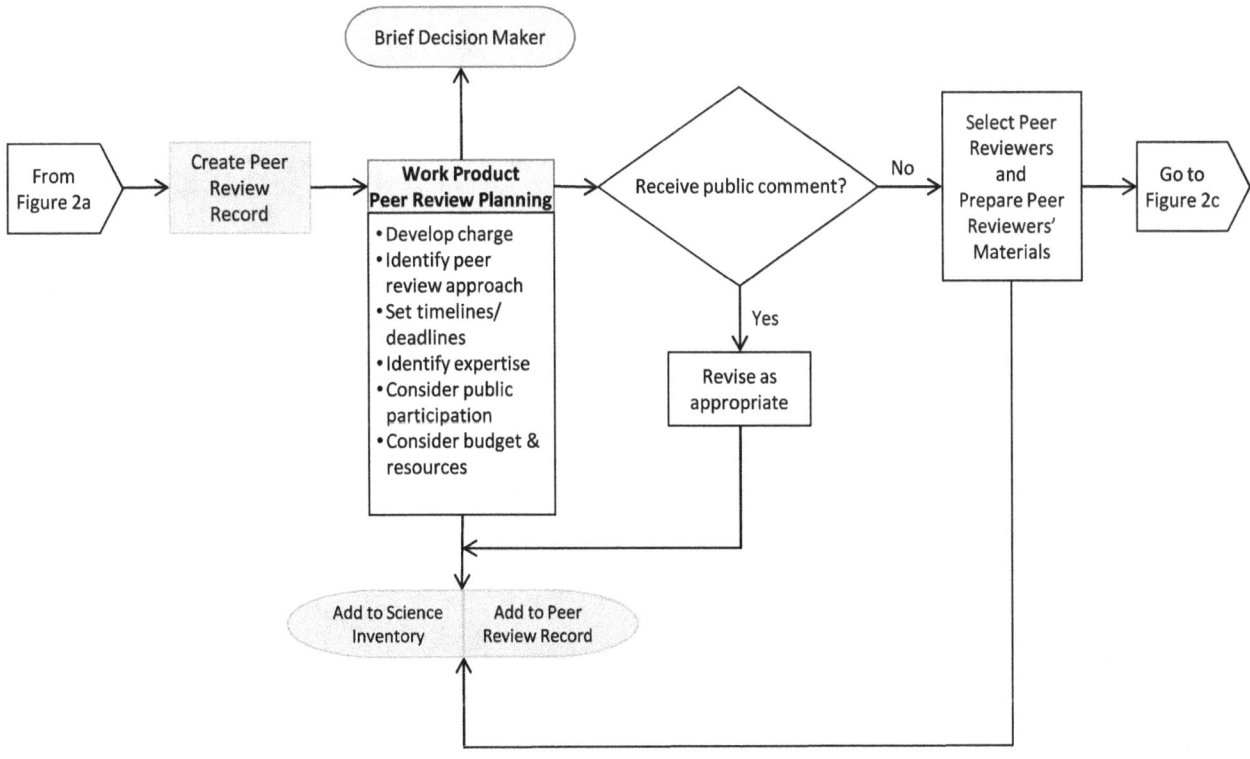

Figure 2b. Planning the Peer Review for Influential Scientific Information (Including HISAs)

1. **If a work product is subject to peer review:**
   - Identify key staff (§ 2.3)
   - Create a peer review record (§ 6.5)
   - Identify criteria/basis for the charge (§ 6.2)
   - Consider options for public participation (§ 7.2)
2. **Develop the draft charge (§ 6.2):**
   - Determine which key issues to address
   - Add to the SI and peer review record
3. **Ensure adequate resources for the peer review (§ 1.2.5)**
4. **Identify a peer review approach (§ 4.2):**
   - Internal (§ 4.2.2), external (§ 4.2.3) or both, as appropriate
   - Letter review (§ 4.4):
     - Managed by Agency or contractor (§ 4.6)
   - Panel review (§ 4.5):
     - Managed by contractor or federal advisory committee (FAC) (§§ 4.6, 4.7)
     - One-time or multiple meetings (§§ 1.2.3, 4.2.1)
   - Add to the SI and peer review record
5. **Set timelines/deadlines:**
   - When will the review be started?
   - What are the intermediate checkpoints?
   - What is the deadline for completion?
   - Add to the SI and peer review record

6. **Identify expertise (§ 5):**
   - Determine the expertise needed (§§ 5.2.1, 5.2.4)
   - Determine sources of peer reviewers (§ 5.2.2)
   - Consider asking the public to nominate peer reviewers (§ 5.2.2)
   - Consider and address the balance of the panel (§ 5.2.4)
   - Consider COIs (§§ 4.6.4, 5.3)
   - Particularly for a HISA, evaluate rotation (§ 5.2.8)
   - If a contractor-managed panel peer review, note special considerations (§ 4.6.4)
   - Formalize arrangement with peer reviewers
   - Add to the SI and peer review record
7. **Determine whether, on what and when public may provide comment (e.g., work product, charge, peer reviewers) (§ 7.2):**
   - Revise peer review plan accordingly
   - Document in the SI and peer review record
   - If a HISA, include a public comment process as part of the peer review whenever feasible and appropriate
8. **Prepare materials for the peer review (§ 6.2.5):**
   - Obtain materials from the Project Manager
   - Prepare instructions for peer reviewer (§ 6.2.5)
   - Include a copy of materials in the peer review record (§ 6.5.2)

Note: Some of these steps may occur concurrently.

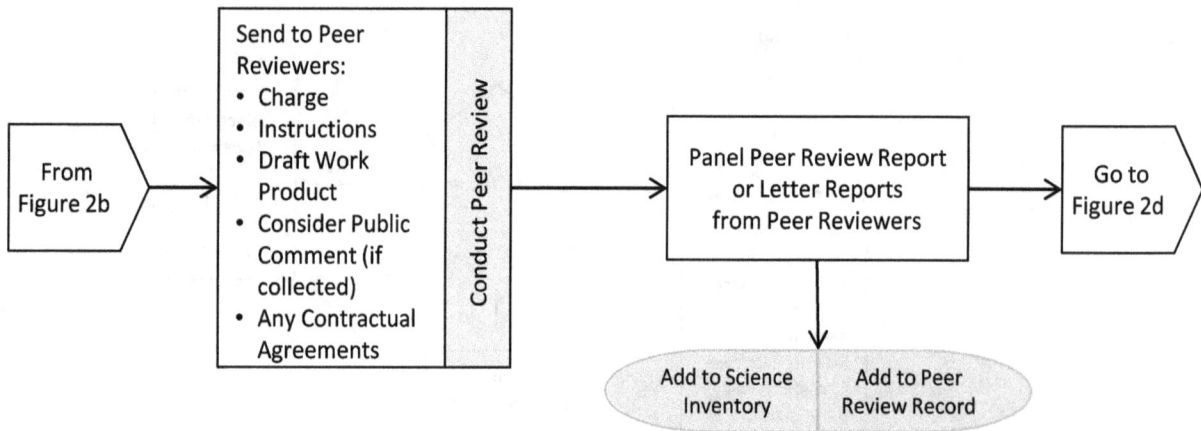

Figure 2c. Conducting the Peer Review of Influential Scientific Information (Including HISAs)

1. **Provide materials to the peer reviewers (§ 6.2.5):**
   - ➤ Charge
   - ➤ Instructions
   - ➤ Draft work product
   - ➤ Public comments if plan provided for public comment on work product
   - ➤ Any contractual agreements associated with the review
   - ➤ Particularly for HISAs, supporting materials for key decisions and findings
2. **Conduct the peer review:**
   - ➤ Particularly if a HISA, public may present comments to peer reviewers at a panel meeting (should be part of peer review plan)
3. **Ask reviewers to prepare peer review comments (§ 6.2.5)**
4. **Prepare Peer Review Report** (collective comments from peer reviewers) (§ 6.2.5)
   - ➤ If conducted by a panel, receive panel peer review report
   - ➤ If conducted by letter, receive individual letter reviews and prepare consolidated peer review report
5. **Add peer review report to the SI and peer review record**

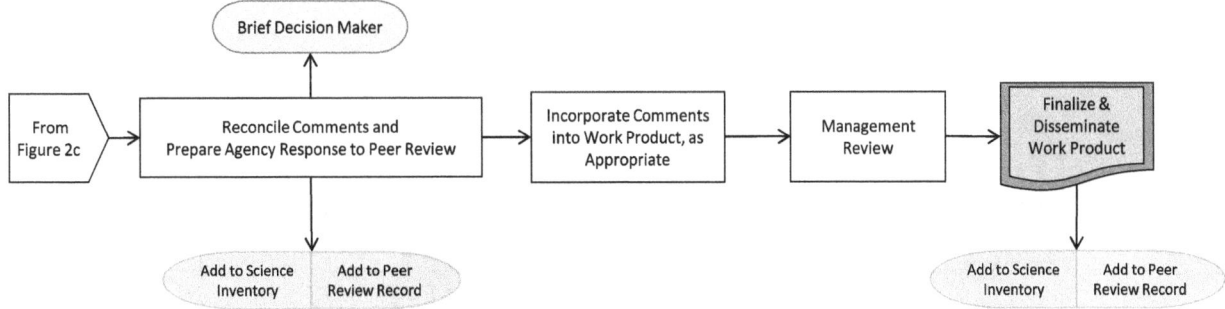

Figure 2d. Completing the Peer Review of Influential Scientific Information (Including HISAs)

1. Evaluate comments from peer reviewers:
  ➤ Consider comments
  ➤ Obtain clarification, if needed
  ➤ Include comments in peer review record
2. Brief the DM on proposed reconciliation of comments
3. Reconcile comments:
  ➤ Revise the work product by incorporating comments, as appropriate
  ➤ For a HISA, prepare a written Agency response and document why any comments were not used
  ➤ Include documentation in peer review record
4. Finalize work product:
  ➤ Include in peer review record
  ➤ Post peer review report and related materials (e.g., charge, Agency response) on the Internet through the SI:
    • For an ISI, post written Agency response to the peer review report, if prepared
    • For a HISA, post written Agency response to the peer review report
  ➤ For all ISI/HISAs that support rulemaking:
    • Include peer review discussion and certification in preamble of the rule

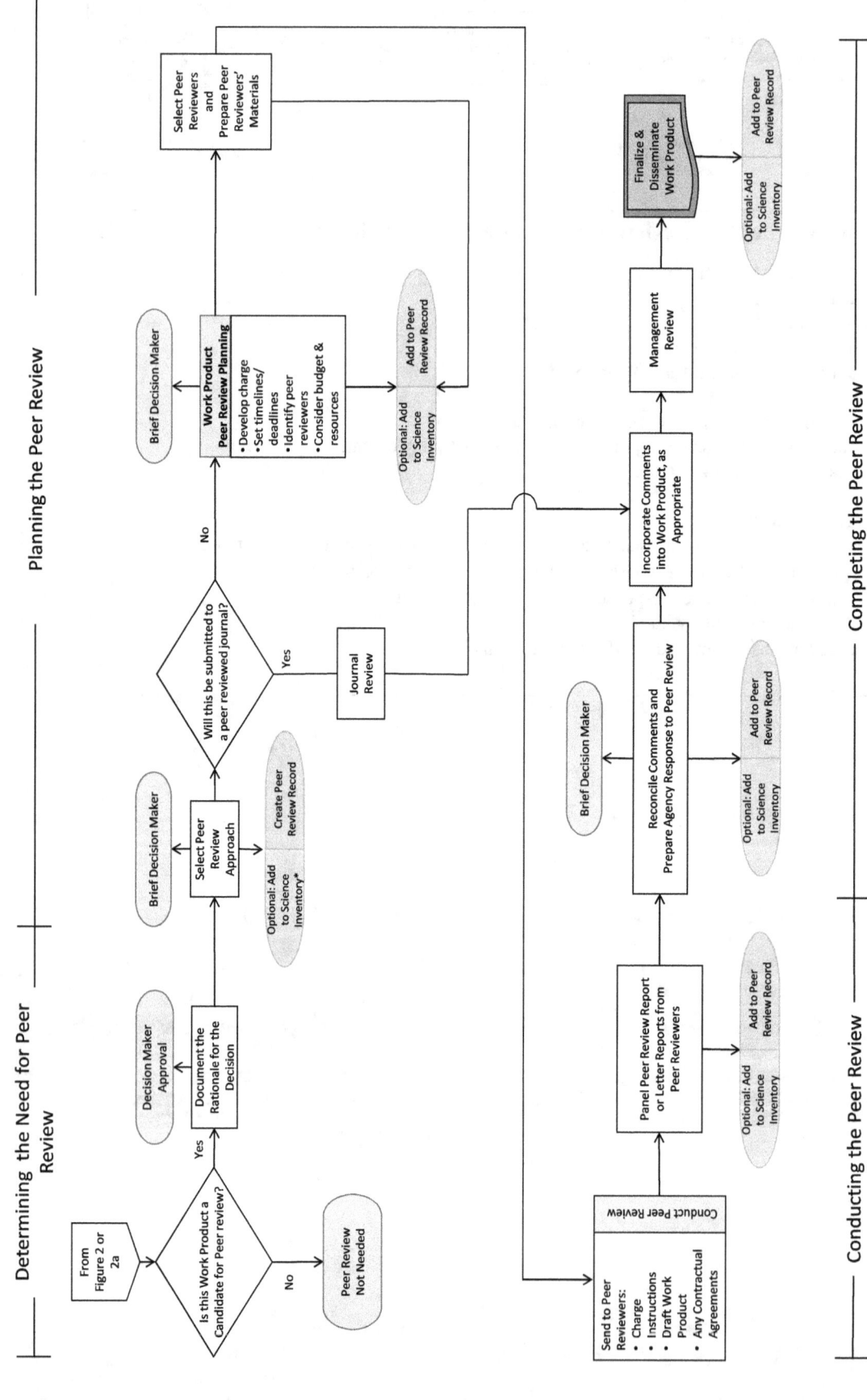

Figure 3. Detailed Flowchart for Other Work Products

* Although other Work Products (non-Influential) may be added to the Science Inventory, they will not be on the EPA Peer Review Agenda

EPA Peer Review Handbook: Roadmap

R.3. Organizing the Peer Review Process

R.3.1. Planning the Peer Review

Planning a peer review is a critical first step to ensuring a successful peer review of a work product. The initial step is to determine whether the work product (either at the conceptual stage or while under development) should be peer reviewed. Once it has been determined that a peer review will be conducted, the DM and Peer Review Leader (PRL) need to plan an appropriate review. This includes:

- categorizing the work product and documenting the decision for influential work products;
- determining resources (budget and personnel);
- scheduling for completion of the peer review;
- creating the peer review record;
- making decisions about an appropriate peer review approach, which considers the forum (i.e., internal and/or external), type (i.e., letter or panel) and mechanism for conducting the review (i.e., Agency-managed, contractor-managed, Federal Advisory Committee [FAC], National Academy of Sciences [NAS]);
- planning for opportunities for public participation;
- developing the charge;
- selecting peer reviewers; and
- preparing materials for the reviewers.

Conceptualizing the Peer Review, which includes defining roles, responsibilities and resources, should take place at the very earliest stages of a product's development. Resources, including personnel, time and funding, should be considered. Based on individual EPA office procedures, other considerations might include the need for briefings, quality assurance (QA) components and reviews and pre-dissemination review planning and approvals.

Categorizing the Work Product (Figure 2a) is based on objective criteria associated with whether the work product is considered influential (i.e., is categorized as ISI), and if influential, whether it is a HISA.

Planning the Peer Review for Influential Scientific Information (Including HISAs) (Figure 2b) takes into account the work product categorization in determining the forum, type and mechanism of peer review. Evaluation and selection of peer reviewers are also documented in the plan, as well as decisions about public participation, preparation of the charge, instructions to reviewers and other information that may be useful to reviewers. For HISAs, in particular, it is important to include sufficient information, including background information about key studies or models, to enable reviewers to understand how significant findings or conclusions in the draft assessment were made.

The charge should be drafted before selection of the peer reviewers to ensure that they have the appropriate expertise to address the questions raised. Developing and maintaining a peer review record should begin at the planning stage of the peer review process (see Section 6.5.3).

R.3.2. Conducting the Peer Review

The success and usefulness of any peer review depends on the quality of the draft work product submitted for peer review, the care given to the statement of the issues or "charge," the match between the peer review draft product and the form of peer review, the match between the peer review draft product and the scientific/technical expertise of the reviewers, and Agency use of peer review comments in the final product. In conducting a peer review, each of the foregoing elements requires serious attention.

Figure 2c shows the order of activities for conducting a peer review of a work product categorized as ISI or a HISA. The peer reviewers are expected to prepare and submit peer review reports at the conclusion of their review. For letter reviews, individual reports are submitted; a single report generally is expected from a peer review panel.

R.3.3. Completing the Peer Review and Finalizing the Work Product

Conducting the peer review of the work product is not the final stage of the peer review process. Rather, the peer review process closes with the following major activities: evaluating peer review comments and recommendations, using the peer review comments for completing the final document, completing the peer review record, and including relevant information in the SI (Figure 2d). The final product represents the true end of the peer review process.

R.3.4. Tools for Managing the Peer Review Process

The following Exhibits may be used by EPA offices to plan, track and document decisions associated with peer review. Note that more than one of the following may be needed for a given draft work product:

- The *Regulatory Action Development Checklist for Workgroups* (Exhibit 1) is an aid for those involved in the development of regulatory actions.

- The list of *Recommended Steps for Planning, Conducting and Completing a Peer Review* (Exhibit 2) is to assist the Project Manager (PM) and PRL in tracking the overall peer review process.

- The *Example EPA Peer Review Decision Summary Documentation* (Exhibit 3) is for the DM, Peer Review Coordinator (PRC) and PRL to document decisions, including the work product categorization, mechanism of peer review and public participation.

Tools and products to enhance the transparency and reporting of peer reviews are summarized in Table 1.

Exhibit 1. Regulatory Action Development Checklist for Workgroups

This checklist will help workgroups plan for peer review in the larger context of regulatory development. Each numbered section corresponds to a time period in the regulatory development process.

1. **Peer Review Prior to Proposal**
   Tier 1 or Tier 2 Rule*
   ___ Is the peer review schedule incorporated into the analytic blueprint?
   ___ Does this rule rely upon influential scientific information (ISI/HISA)?
   ___ Will the work product be reviewed using external peer review?
   Tier 3 Rule
   ___ Is the peer review schedule incorporated into the plans for producing the action?
   ___ Does this rule rely upon ISI or a HISA?
   ___ If an internal mechanism will be used for peer review, is it acceptable according to the *Peer Review Handbook*?

2. **Sending a Proposed Rule Forward for the Administrator's Signature**
   ___ Has peer review been completed?
   ___ Does the action memorandum indicate whether the rule relies upon ISI or a HISA?
   ___ If the proposed rule relies on ISI or a HISA, is there a discussion of the peer review in the preamble of the rule?

3. **Before the Proposed Rule Publishes**
   ___ Were the peer review report and any relevant materials included in the docket for this rulemaking?

4. **Peer Review Prior to Finalization**
   ___ Is a new peer review plan necessary as a result of new regulatory options?

5. **Sending a Final Rule Forward for the Administrator's Signature**
   ___ Has any new peer review of the work product been completed?
   ___ Does the action memorandum indicate whether the rule relies on ISI or a HISA?
   ___ If the final rule relies on ISI or a HISA, is there a discussion of the peer review in the preamble of the rule?

6. **Before the Final Rule Publishes**
   ___ Were the peer review report and any relevant materials included in the docket for this rulemaking?

Note: For ISI and HISAs, the administrative record for the action should include a certification explaining that the action is consistent with provisions of the Office of Management and Budget (OMB) Peer Review Bulletin (see Appendix C).

*For further information on tiering and criteria used to determine the appropriate tier for an action, see http://intranet.epa.gov/actiondp/adp-milestones/tiering.htm .

Exhibit 2. Recommended Steps for Planning, Conducting and Completing a Peer Review

Recommended Steps	Comments
I. **Categorize the work product and document your rationale (requires Decision Maker [DM] approval)** (see *Example EPA Peer Review Decision Summary Documentation* form and Chapter 3) __ Influential scientific information (ISI) __ Highly influential scientific assessment (HISA) __ Other	

II. **Plan the peer review and brief the DM (Chapters 4 and 5)**
 __ Begin creating a peer review record
 __ Select the peer review approach
   • Internal, external or both
   • Letter or panel
   • EPA- or contractor-managed
 __ Set timelines/deadlines
 __ Consider budget and resources
 __ Develop charge questions
 __ Identify areas of expertise needed
 __ Consider public participation, stakeholder involvement
 __ Identify and evaluate potential peer reviewers (expertise and ethics issues)
 __ For HISAs and ISI, create public peer review plan and add other relevant information in the EPA Science Inventory * (see Chapter 7)
 __ Formalize arrangements with the selected peer reviewers	

III. **Conduct the peer review (Chapter 6)** __ Send peer review materials (e.g., charge and instructions, draft work product and supporting materials, contractual agreements, public comments) to peer reviewers __ Convene panel or conduct letter review __ Obtain reviewers' comments (peer review report)	
IV. **Complete the peer review and brief the DM (Chapters 6 and 7)** __ Reconcile reviewers' comments and document how comments were addressed __ Finalize work product __ Update peer review record __ For HISAs and ISI, post the peer review report, any Agency response (necessary for a HISA), and the final work product	

* EPA. Peer Review Agenda. http://cfpub.epa.gov/si/si_public_pr_agenda.cfm.

Exhibit 3. Example EPA Peer Review Decision Summary Documentation

1) WORK PRODUCT TITLE:

_____

2) WORK PRODUCT DESCRIPTION:

_____

3) Assistant Administrator (AA)-ship or Region and Originating Office/Division:

_____

4) Decision/Rule/Regulation/Action/Activity That the Work Product Supports: _____

5) Categorization of Work Product (see page 2 of this exhibit for explanation):
   __ Influential Scientific Information (ISI)
   __ Highly Influential Scientific Assessment (HISA)
   __ Other Scientific or Technical Work Product

6) Rationale for Work Product Categorization and if Peer Review is needed: _____

7) Peer Review Mechanism(s) to Be Used, If Applicable (check all that apply):
(If the work product is designated as ISI or a HISA, conduct peer review [unless exempted or deferred]. For other scientific or technical work products, peer review should be conducted if the Decision Maker [DM] determines that it is appropriate. Evaluate and allot sufficient resources, including funds, time and personnel.)

   __ Peer Review Not Necessary (provide rationale)
   __ Internal
   __ External: Submit to Peer-Reviewed Journal
   __ External: Letter Reviews

   __ External: Contractor-Managed Panel
   __ External: Federal Advisory Committee (FAC) (e.g., Science Advisory Board [SAB])
   __ External: Other Panels (e.g., National Academy of Sciences [NAS])

8) Opportunities for Public Participation (check all that apply):
   __ Comment on Charge
   __ Nominate Potential Peer Reviewers
   __ Comment on Potential Peer Reviewers

   __ Comment on Draft Work Product
   __ Comment on Peer Review Mechanism
   __ Oral Presentation to Reviewers

Documentation/Approval of Decision for an ISI or HISA Work Product
Peer Review Leader (Recommendation) _____ Date _____
Peer Review Coordinator (Concurrence) _____ Date _____
Decision Maker (Approval) _____ Date _____

The DM must approve the categorization decision for work products designated as ISI or HISA. Work products designated as ISI or HISA should be peer reviewed; for HISA, external peer review is the approach of choice. For work products not designated as ISI or a HISA, peer review should be conducted if the DM determines it is appropriate.

If the ISI/HISA work product is exempted or deferred from peer review, state the reason(s) why:

Note: Exemption or deferral from peer review of an ISI or HISA requires Administrator approval.

Exhibit 3. Example EPA Peer Review Decision Summary Documentation: Explanation

Yes/ No	Item/Instructions	Handbook Section
	**Designate the Work Product Category*— DM and Peer Review Coordinator (PRC)**	
	Is Work Product Scientific or Technical (includes economic and social science work products)?	3.1.1
	**If scientific or technical, which designation does the work product best fit:**	
	ISI:[†] Will have or does have a clear and substantial impact on important public policies or private sector decisions. Decision Makers should consider the following factors when determining whether a product is likely to be influential: • Establishes a significant precedent, model or methodology. • Is likely to have an annual effect on the economy of $100 million or more. • Is likely to adversely affect in a material way the economy; a sector of the economy; productivity; competition; jobs; the environment; public health or safety; or state, tribal or local governments or communities. • Addresses significant controversial issues. • Focuses on significant emerging issues. • Has significant cross-Agency/interagency implications. • Involves a significant investment of Agency resources. • Considers an innovative approach for a previously defined problem/process/methodology. • Satisfies a statutory or other legal mandate for peer review.	3.2.1
	HISA: A scientific assessment (i.e., an evaluation of a body of scientific/technical knowledge that typically synthesizes multiple inputs, data, models and assumptions and/or applies best professional judgment to bridge uncertainties in available information) that meets the following: • In addition to meeting the criteria for ISI, could have a potential impact of more than $500 million in any year; or • Is novel, controversial or precedent-setting or has significant interagency interest.	3.2.3
	Other (includes journal articles): • Define in comments.	3.2.5

* Designation of a work product's category could change during the course of development. Any changes in designation also should be documented and approved (see Section 3.2.7).

[†] For examples of Agency work products designated as ISI and HISAs, see the Peer Review Agenda website (http://cfpub.epa.gov/si/si_public_pr_agenda.cfm).

Table 1. Agency Tools and Products for Peer Review Transparency and Reporting

Tool (T)/Product (P)	Description	Handbook Section
(T) Roadmap Flowcharts	Graphically describe the Agency's peer review process.	Roadmap
(T) Example Decision Summary Documentation	Individual product documentation is used in each EPA office to start a record of management decision and approval to categorize a product and the type of peer review it will undergo. This document is used at the EPA office level.	Roadmap Exhibit 2
(T) Conducting a Peer Review	A planning and implementation tool for anyone managing the peer review process of a work product.	Roadmap Exhibit 1
(P) Public Peer Review Plan (automatically generated in the SI when information on ISI or a HISA is entered). The SI is a tool to help generate the public peer review plan.	Begin a systematic process of peer review planning for ISI and HISAs that an Agency plans to disseminate in the foreseeable future. Each peer review plan includes: • A paragraph including the title, subject and purpose of the planned report, as well as an Agency contact to whom inquiries may be directed to learn the specifics of the plan. • Whether the dissemination is likely to be ISI or a HISA. • The timing of the review (including deferrals). • Whether the review is conducted through a panel or individual letters (or whether an alternative procedure is exercised). • Whether there are opportunities for the public to comment on the work product to be peer reviewed, and if so, how and when these opportunities are provided. • Whether the Agency provides significant and relevant public comments to the peer reviewers before they conduct their review. • The anticipated number of reviewers (3 or fewer, 4–10 or more than 10). • A succinct description of the primary disciplines or expertise needed in the review. • Whether reviewers are selected by the Agency or by a designated outside organization. • Whether the public, including scientific or professional societies, are asked to nominate potential peer reviewers.	7.3.4
(P) Peer Review Charge	As part of each peer review, the PRL formulates a clear, focused charge that identifies the technical and scientific issues on which the Agency would like feedback and invites suggestions for improving the document as a whole. This request signals the Agency's receptivity to expert recommendations. The charge to peer reviewers usually makes two general requests. First, it focuses the review by presenting specific questions and concerns surrounding such issues as the comprehensiveness of the literature reviewed, the soundness of the method used, the scientific support for the assumptions employed, and the sensitivity analysis (i.e., the sensitivity of the results to alternative assumptions). Secondly, it invites general comments on the work product as a whole.	6.2
(P) The Peer Review Report(collective comments from peer reviewers)	The collective comments on the scientific or technical work product undergoing peer review provided by the peer reviewers in response to the peer review charge is called the Peer Review Report. The EPA makes the reports for ISI and HISAs available on the SI website, which links directly to the Peer Review Agenda entry for that item.	6.2.5

Table 1. Agency Tools and Products for Peer Review Transparency and Reporting

Tool (T)/Product (P)	Description	Handbook Section
(P) Agency's Response to Peer Review Report	The PRL should evaluate and analyze all peer review comments and recommendations carefully. The peer review of a work product is not complete until the peer review comments are incorporated into the final version or reasons are stated why such comments are not incorporated. The peer review record is complete only when it contains a copy of the final work product (when there is one) that addresses the peer review comments and a copy of the response-to-comments document. The PRL should brief the DM on how to address the peer review comments. Per the OMB Peer Review Bulletin, the Agency's response to the peer review report for HISAs should be posted on the SI.	6.3
(P) Peer Review Record	The peer review record is the formal record (file) of decision on the conduct of the peer review, including the type of peer review performed and an explanation of how the peer review comments are addressed. It includes sufficient documentation for an uninvolved individual to understand what happened and why. The peer review record is separate from the entry in the SI. Although some information from the peer review record appears in the SI, the paper peer review record is the official record of the peer review. The PRL (with the Project Manager [PM], if there is one) creates a separate, clearly marked peer review file within the overall file for development of the work. Once the peer review is completed, it is the responsibility of the PRL to ensure that the peer review record is filed and maintained in accordance with the organization's document retention procedures.	6.5
(T) Science Inventory	The SI (www.epa.gov/si) is a searchable database that contains information on EPA publications and presentations. The SI is used to track the Agency's work products that are categorized as ISI and HISAs, including their status and peer review plans. EPA offices are expected to keep this information current by updating SI entries for ISI and HISAs at least every 6 months.	7.3.1, 7.3.2, 7.3.3
(P) Peer Review Agenda	The Peer Review Agenda (PRA) is a component of the EPA SI. ISI and HISA work product metadata, including peer review information and related documents, are entered into the SI and then published to the Agency PRA, which informs EPA website visitors about EPA's planned and ongoing peer review activities. The website for the EPA's Peer Review Agenda is http://cfpub.epa.gov/si/si_public_pr_agenda.cfm.	7.3.3
(P) Annual Report on Peer Review to OMB	Consistent with the OMB's Peer Review Bulletin, the EPA expects to submit a report to OMB each year. This report includes information concerning the peer reviews conducted on ISI and HISAs during the previous fiscal year. The EPA generates this report from the information in the SI.	7.4

PEER REVIEW GUIDANCE

1. Peer Review at EPA: General Concepts and Context

1.1. Overview

> Peer review of all scientific and technical information that is intended to inform or support Agency decisions is encouraged and expected. Influential scientific information, including highly influential scientific assessments, should be peer reviewed in accordance with the Agency's *Peer Review Handbook*. All Agency managers are accountable for ensuring that Agency policy and guidance are appropriately applied in determining if their work products are influential or highly influential, and for deciding the nature, scope, and timing of their peer review. For highly influential scientific assessments, external peer review is the expected procedure. For influential scientific information intended to support important decisions, or for work products that have special importance in their own right, external peer review is the approach of choice. Peer review is not restricted to the nearly final version of work products; in fact, peer review at the planning stage can often be extremely beneficial.
>
> —EPA Peer Review Policy Statement, 2006

To implement the EPA's Peer Review Policy (Appendix A) effectively, individuals involved in peer review activities need to understand what peer review is and why the Agency conducts peer reviews. Those individuals also need to understand how peer review differs from activities such as peer input, stakeholder input and public comment. Familiarity with federal and EPA guidelines related to peer review is essential. This chapter discusses each of these topics and also addresses the role of peer review in regulatory development.

1.2. Peer Review

1.2.1. What Is Peer Review?

Peer review is a documented process for enhancing a scientific or technical work product so that the decision or position taken by the Agency, based on that product, has a sound, credible basis. (For a discussion of what constitutes a scientific or technical work product, see Section 3.1.1.) It is conducted by qualified individuals

> *The goal of peer review is to obtain an independent review of the product from experts who have not contributed to its development.*

(or organizations) who are independent of those who performed the work and who are collectively equivalent in technical expertise to those who performed the original work (i.e., peers). Peer review is conducted to ensure that activities are technically defensible, competently performed, properly documented and consistent with established quality criteria. Peer review is an in-depth assessment of the assumptions, calculations, extrapolations, alternate interpretations, methodology, acceptance criteria and conclusions pertaining to the scientific or technical work product, and of the documentation that supports them. Peer review also may provide an evaluation of a topic where quantitative methods of analysis or measures of success are unavailable or undefined. Peer review usually is characterized by a one-time or limited number of interactions by independent peer reviewers who provide responses to a series of questions included in a "charge" developed by EPA (see Section 6.2.1). Peer review is encouraged during the development of a project or method, and/or as part of the culmination of the work product, as appropriate. Regardless of the timing of peer review, the goal is to ensure that the final product is scientifically and technically sound.

1.2.2. Why Use Peer Review?

Peer review is intended to identify any technical problems or unresolved issues in a preliminary (or draft) work product through the use of independent experts. This information then is used to revise the draft product so that the final work product will reflect sound scientific and technical information and analyses. To be most effective, <u>peer review of a scientific or technical work product should be incorporated into the up-front planning of any action based on the work product; this includes obtaining the proper resource commitments (personnel and money) and establishing realistic schedules</u>.

Although conducting a peer review requires an up-front commitment of time and resources, the benefits usually justify these added resources. Peer review enhances the credibility and acceptance of the decision based on the work product. Also, by ensuring a sound basis for decisions, cost savings are likely to be realized because decisions are less likely to be challenged.

> *Peer review is not free; however, not doing peer review can be costly.*

1.2.3. When and How Often Should Peer Review Occur?

The Agency has significant discretion in deciding on the timing and the frequency of peer review. Options abound, each with merits depending on the context and specified peer review objectives. In many situations, a single peer review event, beginning when the final draft work product becomes available, is the approach taken. It is increasingly apparent, however, that peer review performed earlier in the work product development stages can provide a superior approach for some work products. There may be substantial incremental benefit to conducting more than one peer review during work product development, particularly when development involves complex tasks, has decision branching points, or could be expected to produce controversial findings. Sometimes additional peer reviews are conducted if the product changes significantly after the initial peer review, or if the Agency would like to know whether the peer reviewers' comments were adequately addressed in the revised product. In addition, early review could be beneficial at the stage of research design or data collection planning when the product involves extensive primary data collection. The Decision Maker (DM) should determine when the peer review(s) should occur, considering the type of work product under development and at what point a peer review would be most beneficial (see Sections 2.3.2 and 3.1.3).

Other types of work products that could benefit from early, up-front peer review in their development include scientific and technical planning products. Examples of such products are research proposals, plans and strategies. Although more than one peer review can be beneficial, the distinction between peer input and peer review should be kept in mind. Experts providing input during the development or planning stages of the work product generally do not become peer reviewers of that product. For more on this distinction, see Sections 1.2.11 and 5.2.7.

1.2.4. What Factors Are Considered in Setting the Timeframe for Peer Review?

The peer review schedule is a critical feature of the process. The schedule should take into account the availability of a quality draft work product; deadlines for the completion of a project, research program or rulemaking; funding availability; availability of qualified peer reviewers; the complexity and length of the product; the possible need to seek public comment on the peer review product; statutory and/or court-ordered deadlines; and logistical aspects of the peer review (e.g., contracting procedures).

The time required to complete an external peer review will depend greatly on the peer review mechanism selected, ranging from several months for individual letter reviews to 10 to 12 months for a review by a federal advisory committee (FAC) *ad hoc* panel or more than a year for a review by a National Academy of Sciences (NAS) panel. Federal Advisory Committee Act (FACA) requirements for advanced notification of committee meetings and opportunities for public participation add to the time required to complete the review but enhance the transparency of the peer review process. Regardless of the peer review mechanism selected, the schedule must include adequate time to evaluate prospective peer reviewers for ethics issues such as potential conflicts of interest (COIs) or an appearance of a loss of impartiality (see Section 5.3).

1.2.5. What Budgetary Factors Should Be Considered in Planning a Peer Review?

Resources necessary to perform peer review should be requested as part of the costs of projects, rules or guidance. For purposes of budget planning, the costs of peer review would include the allocation of staff resources (full-time equivalents, or FTE), the contract or other costs associated with the use of outside peer reviewers and the administrative costs of conducting a review (e.g., copying, travel expenses). For peer reviews conducted by the Science Advisory Board (SAB) or Clean Air Scientific Advisory Committee (CASAC), the SAB Staff Office budgets for the peer review, including peer reviewer travel expenses, contract costs for meeting support and FTEs to support the advisory committee's work.

> *Peer review is part of the normal cost of doing business.*

Senior management in EPA offices should ensure that budget requests include anticipated resources for peer review. (It should be noted that the term "EPA offices" in this Handbook refers to all headquarters, regional and program offices.) Peer review should be considered as a normal part of doing business. Peer review resource considerations also should be addressed in the analytic blueprint for Agency rulemaking actions.

1.2.6. Who Are the Peer Reviewers?

Peer reviewers are individuals who have technical expertise in the subject matter of the work product undergoing peer review. For this reason, they may be referred to as "subject matter experts." Peer reviewers should not be associated with generating the work product undergoing review; they should be able to offer independent scientific advice. Peer reviewers need to be willing participants in the peer review process; they should agree to read all materials, participate fully and act ethically. Peer reviewers should maintain the confidentiality of the product and information contained in the product (when necessary), perform the review within the agreed-upon timeframe and be unbiased and objective. Peer reviewers should disclose any activities or circumstances that could pose a conflict of interest or create an appearance of a loss of impartiality that could interfere with an objective review. See Chapter 5 for a thorough discussion of peer reviewer qualifications and ethical considerations.

1.2.7. What Is the Difference Between Internal and External Peer Review?

An internal peer review is a technical or scientific review by individuals from within the Agency who have the appropriate expertise and are independent from the development of the work product. Internal peer reviewers should come from a different organizational unit than the one in which the work originates. Examples of internal peer review mechanisms may be found in Section 4.2.2.

An external peer review is a review by non-EPA experts with appropriate knowledge and skills who are independent from the development of the work product. External reviewers may come from other federal agencies, state and local government agencies, academia, industry, nongovernmental organizations or other outside organizations. Examples of external peer review mechanisms may be found in Section 4.2.3.

For work products that are intended to support important public policy or private sector decisions, external peer review is the approach of choice. Note that an internal peer review or technical review often precedes an external peer review. Refer to Section 4.2.1 for guidance on when to use internal and external peer reviews.

1.2.8. What Is the Difference Between Internal Peer Review and Internal Management Review?

An internal peer review is an assessment of the scientific and technical quality of a work product by independent Agency experts prior to the publication or release of the work product outside the Agency. An internal management review (sometimes referred to as "clearance") is a process for obtaining line management approvals prior to the work product's release or publication. While an internal peer review may be included as part of the internal management review (as in the case of a technical review conducted prior to the submission of a manuscript to a journal), the internal management review does not substitute for an internal peer review.

1.2.9. What Is a Letter Peer Review?

A letter review takes place when EPA seeks individual written peer review comments from independent experts, typically in the form of correspondence to EPA from the peer reviewer. The number of reviewers selected depends largely on the scientific and technical expertise required to address the issues presented in the peer review charge. Each reviewer evaluates the draft technical work product independently without consultation with other reviewers. No collaborative or consensus peer review report is developed. For letter reviews managed by a contractor, the contractor may compile all peer review comments into a single report but should not edit the comments in any way, transmitting comments unaltered to EPA. For more information on letter peer reviews, see Section 4.4.

1.2.10. What Is a Peer Review Panel?

A peer review panel is a group of experts who share and discuss their peer review comments with one another, regardless of whether the sharing takes place in a face-to-face meeting or via email or teleconference. The number of panel members selected for a peer review will depend on the issue being investigated, the time available and resources. Individuals should have appropriate scientific and technical expertise such that the review panel as a whole covers the broad spectrum of expertise necessary to address the issues and questions presented in the peer review charge. For some panels, members may be asked to prepare individual comments for submission to the Agency; for others, the panel members may be asked to collaborate and provide consensus advice in a single report to EPA. If panels provide collective or consensus (rather than individual) advice, they may be subject to the requirements of the FACA, which imposes certain open meeting, balanced membership and committee chartering requirements. For more information on peer review panels, including FACs, see Chapter 4.

1.2.11. What Is Peer Input, and How Does It Differ From Peer Review?

Peer input, sometimes referred to as peer consultation, is a form of peer involvement that generally connotes an interaction during the development of an evolving Agency work product, providing an open exchange of data, insights and ideas. Such input may be continued and iterative, and it often involves scientific and technical experts from both inside and outside the Agency. A common example is the input received from workgroup members during the development of a product.

> *Peer input is not a substitute for peer review.*

The key distinctions between peer input and formal peer review are the independence of the peer reviewers and their level of involvement. Generally, someone who provided peer input on a work product no longer is considered independent and should not become a peer reviewer for that same work product.

Peer input provides valuable contributions to the development of the work product. Peer input does not substitute, however, for peer review. In other words, one cannot argue that a peer review is not necessary simply because a work product has received "enough" peer input.

1.2.12. What Is Stakeholder Involvement, and How Does It Differ From Peer Review?

Stakeholder involvement occurs when the Agency engages a select set of individuals, groups or representatives from organizations or interest groups that have a stake in the outcome of the EPA's work and policies or that seek to influence the Agency's future direction to work directly on specific issues.

> *Stakeholder involvement is not a peer review mechanism.*

The Agency often seeks stakeholder involvement to ensure that all relevant facts and viewpoints related to the issue are considered. This is an interactive process that usually involves other agencies, industry groups, regulated-community experts, environmental groups and other interest groups that represent a broad spectrum of the regulated community, among others. The process of stakeholder involvement usually strives for general agreement among the involved groups and may be subject to the FACA. Stakeholders should not be involved in the peer review process if there has been prior engagement with the Agency on the development of the product or the issue. If stakeholders are involved in the peer review process, they must meet all applicable ethics laws and regulations.

Although stakeholder involvement is an outreach activity that contributes greatly to the development of a work product, it is not considered a peer review mechanism.

1.2.13. How Does Public Comment Differ From Peer Review?

The critical distinction between public comment and peer review is that public comment does not necessarily draw the kind of independent, expert information and in-depth analyses expected from the peer review process. Public comment frequently is open to all issues, and may be solicited for policy purposes or as part of the regulatory process, whereas the peer review process focuses on scientific and technical issues specified in the peer review charge.

Public comment solicited from the general public through the *Federal Register* or by other means may be required by the Administrative Procedure Act or other statutes. Public commenters usually include a broad array of individuals; some may be scientific experts (and may provide peer input), some may be experts in other areas, and some are interested non-experts.

In terms of peer review, public comments can provide important input to the identification and selection of peer reviewers, the refinement of charge questions to be addressed in peer review, and identification of technical issues to be considered by the peer reviewers. Generally, public comment enhances the transparency of the peer review process. Although it may be an important component of the EPA's decision-making process, public comment does not substitute for peer review. See Section 7.2 for more information on public participation in the peer review process.

1.3. Policies and Guidance That Relate to Peer Review

To provide the framework for ensuring the credibility and utility of the Agency's science, EPA relies on its Peer Review Policy and peer review procedures and guidelines in this *Peer Review Handbook*; guidance from the Office of Management and Budget (OMB) Peer Review Bulletin; and the EPA's Quality System, *Information Quality Guidelines* and Scientific Integrity Policy. Each is briefly discussed below.

1.3.1. What Is the EPA's Peer Review Policy?

The EPA's Peer Review Policy[3] was first issued in 1993 and was updated in 2006 (see Appendix A). It emphasizes the critical role of peer review in ensuring that the EPA's decisions rest on sound science and data.

1.3.2. What Are the Legal Ramifications of the Peer Review Policy?

The Peer Review Policy does not establish or affect legal rights or obligations. Rather, it confirms the importance of peer review where appropriate, outlines relevant principles and identifies factors that Agency staff should consider in implementing the policy. Except where provided otherwise by law, peer review is not a formal part of, or substitute for notice-and-comment rulemaking or adjudicative procedures. The EPA's decision to conduct peer review in any particular case is wholly within the Agency's discretion. Similarly, nothing in the Peer Review Policy creates a legal requirement that EPA respond to peer review comments. To the extent that EPA decisions rely on scientific and technical work products that have been subjected to peer review, however, the remarks of peer reviewers should be included in the record for those decisions.

EPA staff and management should consult with attorney(s) in the Office of General Counsel (OGC) and/or Office of Regional Counsel (ORC), to obtain legal advice related to peer review. OGC has attorneys who are specialists in specific areas (e.g., FACA considerations, contractual responsibilities, ethics issues), and they should be consulted as needed, following consultations with local resources.

---

[3] EPA. 2006. *Peer Review and Peer Involvement at the U.S. Environmental Protection Agency.*
http://epa.gov/peerreview/pdfs/peer%20review%20policy%202006.pdf.

1.3.3. What Is the Office of Management and Budget's Peer Review Bulletin, and How Does It Relate to Peer Review at EPA?

OMB's *Final Information Quality Bulletin for Peer Review*[4] (see Handbook Appendix B), hereafter the OMB Peer Review Bulletin, provides guidance to federal agencies for enhancing the peer review of government science documents and establishes minimum standards for when to conduct peer review. EPA conducts peer review of its products in accordance with the guidance in the OMB Peer Review Bulletin.

OMB's Peer Review Bulletin provides two important definitions:

- **Influential Scientific Information (ISI):** Scientific information that the Agency "reasonably can determine will have or does have a clear and substantial impact on important public policies or private sector decisions."

- **Highly Influential Scientific Assessment (HISA):** A <u>subset</u> of ISI that is a scientific assessment (i.e., an evaluation of a body of scientific or technical knowledge, which typically synthesizes multiple factual inputs, data, models, assumptions and/or applies best professional judgment to bridge uncertainties in the available information) that "could have a potential impact of more than $500 million in any year on either the public or private sector" or "is novel, controversial, or precedent-setting, or has significant interagency interest."

Per the OMB Peer Review Bulletin, all of the Agency's ISI/HISA should be peer reviewed unless they meet specified exemption criteria (see Handbook Section 3.3). Decisions regarding categorization of products as HISA or ISI should be made early in the stages of product development; relevant guidance may be found in Section 4.2.1. The OMB Peer Review Bulletin instructs federal agencies to establish a process for public disclosure of peer review planning, including a Web-accessible description of the plan that each agency has developed for reviewing its ISI and HISAs. An agenda of the Agency's plans for reviewing these products may be found on the EPA Peer Review Agenda (http://cfpub.epa.gov/si/si_public_pr_agenda.cfm) (see Section 7.3).

1.3.4. What Is the EPA's Quality System, and How Does It Relate to Peer Review?

The Quality System framework consists of policies, procedures and oversight processes that assure the Agency's environmental data are of sufficient quantity and quality to support the data's intended use. All EPA programs generating environmental data and information, or using data and information from non-EPA sources, are to conform to the Agency's Quality Policy, CIO 2105.0 (May 5, 2000)[5], which is based on international quality standards and practices. The EPA Quality System specifies systematic planning for quality and documentation of the data quality requirements for the scientific or technical work product being developed. The Office of Environmental Information has Agency-wide oversight of the mandatory quality system, and the program and regional offices are responsible for developing a Quality Management Plan for implementing their organization-specific Quality Assurance (QA)

---

[4] OMB. 2004. Memorandum for Heads of Departments and Agencies, *Final Information Quality Bulletin for Peer Review*. http://www.whitehouse.gov/sites/default/files/omb/memoranda/fy2005/m05-03.pdf.

[5] EPA. 2000. *Policy and Program Requirements for the Mandatory Agency-Wide Quality System*. EPA Order Classification No. CIO 2105.0. http://intranet.epa.gov/quality/documents/21050.pdf.

program. Each organization has a designated Director of Quality Assurance (DQA) or Quality Assurance Manager (QAM) responsible for quality.

QA and peer review are complementary activities and ensure that EPA uses scientifically sound data and information in making programmatic and regulatory decisions. Peer review does not replace the Agency's mandatory requirements to collect and use data of appropriate quality for the intended use in decision making. QA promotes the application of quality requirements at the project level such as determining precision, accuracy, representativeness, comparability, completeness and sensitivity of the data. Peer review primarily focuses on the scientific soundness of the results and conclusions presented in the work product. It is recognized as a valuable process that provides an objective and transparent assessment of the utility and credibility of the science. QA requirements and activities should be documented during the planning and development of the product prior to peer review. The Handbook encourages the Peer Review Leader (PRL) to contact the organization's quality assurance individual about applicable QA requirements for the product being peer reviewed. QA specifications are usually documented in a Quality Assurance Project Plan.

1.3.5. What Are the EPA's Information Quality Guidelines (IQG), and How Do They Relate to Peer Review?

The EPA's *Guidelines for Ensuring and Maximizing the Quality, Objectivity, Utility, and Integrity of Information Disseminated by the Environmental Protection Agency*,[6] better known as the EPA's Information Quality Guidelines (IQG), contain procedural guidance for ensuring that the information the Agency disseminates to the public is reliable and accurate, appropriate for its intended use, and protected from compromise (i.e., its objectivity, reliability and integrity are maintained). The EPA's IQG allows persons affected by EPA's publicly disseminated information to seek and obtain corrections from EPA (through its Office of Environmental Information). Peer review is a key step in ensuring the quality, objectivity, utility and integrity of the information that EPA disseminates.

> *Products undergoing peer review (pre-disseminated products) need a disclaimer.*

Agency products undergoing peer review are not considered "disseminated" under the EPA's IQG because they are dynamic documents and are subject to change and, therefore, they do not represent the EPA's final decision or position. These "pre-dissemination" products should contain the following disclaimer:

> *This information is distributed solely for the purpose of pre-dissemination peer review under applicable information quality guidelines. It has not been formally disseminated by EPA. It does not represent and should not be construed to represent any Agency determination or policy.*

In cases where the information is highly relevant to specific policy or regulatory deliberations, the disclaimer should appear on each page of the work product. Agency work products that are disseminated after the peer review process is completed are subject to the EPA's IQG.

---

[6] EPA. 2002. *Guidelines for Ensuring and Maximizing the Quality, Objectivity, Utility, and Integrity of Information Disseminated by the Environmental Protection Agency*. EPA/260R-02-008.
http://www.epa.gov/quality/informationguidelines/documents/EPA_InfoQualityGuidelines.pdf.

1.3.6. What Are the General Assessment Factors, and How Do They Relate to Peer Review?

The guidance titled *General Assessment Factors for Evaluating the Quality of Scientific and Technical Information*[7] (see Appendix C) and its addendum[8] complement the EPA's IQG and Quality System and are an additional resource for EPA staff involved in the peer review process. The guidance establishes the EPA's expectations for scientific and technical information that is voluntarily submitted to or gathered by the Agency. Regardless of source, this information must be evaluated for quality and relevance prior to being used in support of EPA actions. The Agency takes into account five general assessment factors to determine whether the information meets its quality requirements: (1) soundness, (2) applicability and utility, (3) clarity and completeness, (4) uncertainty and variability, and (5) evaluation and review. The "evaluation and review" factor refers to the extent of independent verification, validation and peer review of the information. For a previous peer review to be considered adequate by the Agency, it should meet the intent of the EPA's Peer Review Policy, and the rigor of the review should be commensurate with the proposed use of the information by the Agency.

1.3.7. What Is the EPA's Scientific Integrity Policy, and How Does It Relate to Peer Review?

The EPA's *Scientific Integrity Policy*[9] facilitates scientific integrity Agency-wide through: (1) the promotion of scientific and ethical standards; (2) communications with the public; (3) the use of peer review and advisory committees; and (4) professional development. The policy promotes the culture of scientific integrity and enhances transparency within scientific processes.

The policy emphasizes the importance of ensuring that scientific studies used to support regulatory and other policy decisions undergo appropriate levels of independent peer review, and it recognizes the role of FACs (see Section 2.3.6.) in providing transparent, external peer review.

1.4. Peer Review and Regulatory Development

1.4.1. What Role Does Peer Review Have in Regulatory Development?

Peer review of scientific and technical work products that support regulations is an important, fundamental step in policy setting and regulatory development processes. A regulation itself is not subject to the Peer Review Policy. If a regulation is supported by a scientific and technical work product(s), however, that underlying work product(s) should be peer reviewed if it does not meet exemption criteria outlined in Section 3.3.

Sometimes peer review leads to recommendations for new information and analyses that would alter the work product and thus modify the scientific/technical basis for the action or rule it supports. For this reason, a completed peer review is desirable before issuing any regulatory proposal for public comment. If that is not possible logistically because of court or statutory deadlines, or other appropriate reasons,

---

[7] EPA. 2003. *A Summary of General Assessment Factors for Evaluating the Quality of Scientific and Technical Information*. EPA/100/B-03/001. http://www2.epa.gov/sites/production/files/2015-01/documents/assess2.pdf.

[8] EPA. 2012. *Guidance for Evaluating and Documenting the Quality of Existing Scientific and Technical Information. Addendum to A Summary of General Assessment Factors for Evaluating the Quality of Scientific and Technical Information*. http://www2.epa.gov/sites/production/files/2015-01/documents/assess3.pdf.

[9] EPA. 2010. *Scientific Integrity Policy*. http://www.epa.gov/osa/pdfs/epa_scientific_integrity_policy_20120115.pdf.

every effort should be made to complete the peer review before the close of the comment period. Because peer review comments on such work products could be of sufficient magnitude to warrant a revision to the proposed action or rule, every effort should be made to complete the peer review prior to the proposal stage.

1.4.2. What Is the EPA's Action Development Process (ADP), and How Does It Relate to Peer Review?

The EPA's ADP is a process designed to ensure that the Agency develops and issues high-quality rules, policy statements, guidance documents, reports to Congress and other regulatory and non-regulatory actions. It assists the Agency in achieving objectivity and transparency of information. It consists of steps for planning sound scientific and economic analyses to support the action, including peer review of any major scientific or technical work product that supports an Agency action.

1.4.3. How Does the Rulemaking Tier Affect Peer Review?

Tier 1 and Tier 2 rulemakings are, by definition, important Agency rulemakings. Therefore, work products supporting Tier 1 and Tier 2 rules should be scrutinized carefully to determine whether they should undergo peer review. In most cases, scientific and technical work products categorized as ISI or a HISA and supporting a Tier 1 or Tier 2 rulemaking should be externally peer reviewed if they do not meet exemption criteria outlined in Section 3.3.

Work products supporting Tier 3 rulemakings also may benefit from peer review. For work products supporting a Tier 3 rule, both internal and external peer review may be appropriate, depending on the nature of the product and other factors. For more information on the tiering process, see http://intranet.epa.gov/actiondp/documents/adp03-00-11.pdf. For more information on the differences between internal and external peer review, see Section 4.2.

1.4.4. Should Peer Review Be Discussed in the Analytic Blueprint for a Regulation?

Analytic blueprints are a critical part of the EPA's ADP (see Section 1.4.2). A blueprint, which is required for all Tier 1 and Tier 2 actions, spells out a workgroup's plans for the data collection and analyses that will support development of a specific action. The blueprint sets forth how this information will be collected, peer reviewed and used to craft the action within a specific budget and timeframe.

Workgroups should address peer review specifically in each analytic blueprint. For peer review purposes, development of the analytic blueprint is the process whereby the workgroup identifies supporting scientific and technical work products and recommends what kind of peer review is needed. The analytic blueprint should show the schedule of the peer review in the context of the schedule for the overall rulemaking. For more information, see http://intranet.epa.gov/actiondp/documents/adp03-00-11.pdf.

1.4.5. What Role Does Peer Review Have in Regulatory Negotiations?

As with other rules, a negotiated rulemaking itself is not subject to the Peer Review Policy. If the regulatory negotiation is supported by scientific and technical work product(s), however, that underlying work product(s) should be peer reviewed if it does not meet exemption criteria outlined in Section 3.3. This peer review should occur before the negotiation takes place, when possible.

1.4.6. Should the Peer Review Be Discussed in the Preamble of a Regulation?

For proposed and final regulations that rely on ISI and HISAs, the peer review report should be discussed in the preamble, as described in the OMB Peer Review Bulletin. The PRL should take steps to ensure that the rule writer and the regulatory workgroup are aware of this provision of the OMB Peer Review Bulletin. For peer review template language, see Appendix D, Sound Science and Peer Review in Rulemaking.

1.4.7. How Is Peer Review Documented in the Action Memorandum for Regulations?

For all rules requiring the Administrator's signature (proposed and final), the action memorandum should indicate the kind of peer review that took place. The current format for action memoranda accompanying regulatory packages is available at http://intranet.epa.gov/actiondp/adp-templates/index.htm#adp.

2. Peer Review Roles and Responsibilities

2.1. Overview

The roles defined in this chapter provide descriptions of responsibilities of key personnel involved in or conducting peer review at the Agency. These personnel are responsible for ensuring the scientific quality of work products that inform decisions.

The EPA Deputy Administrator (DA) is the senior Agency official for peer review. The DA is ultimately responsible for the performance of peer review for scientific and technical information that is intended to inform and support the EPA's environmental decisions.

The Science and Technology Policy Council (STPC), the Peer Review Advisory Group (PRAG) and the Office of the Science Advisor (OSA) oversee implementation of the Agency's Peer Review Policy. The Office of Research and Development (ORD) is responsible for maintaining the Agency's Peer Review Agenda.[10] EPA Assistant Administrators (AAs) and Regional Administrators (RAs) are responsible for making peer review decisions that are specific to their EPA offices; they may delegate some responsibilities, however, to other Decision Makers (DMs) within their organizations for planning and managing the peer review process in accordance with the Handbook guidelines. The Office of General Counsel (OGC) and Office of Regional Counsel (ORC) provide legal advice to assist Agency personnel in carrying out their peer review-related responsibilities.

Specific roles and responsibilities of agency organizations and personnel associated with peer review are discussed below. EPA employees with assigned peer review responsibilities should be familiar with the Agency's Peer Review policy and receive the appropriate peer review training. The PRAG develops and provides training on the Handbook for all employees with designated peer review responsibilities. See Section 1.2.6 for the roles and responsibilities of the peer reviewer.

> *Employees should be familiar with their roles and responsibilities for peer review.*

2.2. Oversight Responsibilities for the EPA's Peer Review Policy

2.2.1. What Is the Role of the Deputy Administrator?

The DA has the authority to establish Agency-wide peer review policies and guidelines that enhance the credibility of EPA as a scientific agency. The DA is the final arbiter of conflicts and concerns about peer reviews conducted by the Agency.

2.2.2. What Is the Role of the Science and Technology Policy Council?

The STPC (formerly known as the Science Policy Council) is a senior Agency council chaired by the EPA Science Advisor. The STPC identifies critical science and technology policy issues and develops approaches that help advance the Administrator's environmental and public health priorities. The STPC is responsible for overseeing the implementation of the Agency's Peer Review Policy. The STPC meets its peer review responsibilities through oversight of the PRAG.

---

[10] EPA. 2015. *Peer Review Agenda*. http://cfpub.epa.gov/si/si_public_pr_agenda.cfm.

2.2.3. What Is the Role of the Peer Review Advisory Group?

The PRAG assists the STPC in overseeing implementation of the Agency's Peer Review Policy and serves as a technical resource for the Agency. It is a workgroup of representatives from EPA program and regional offices that was established to develop and interpret peer review guidelines, address peer review issues and promote effective peer review practices across EPA. It also serves as a cross-Agency coordination workgroup to increase the quality and consistency of peer reviews at the Agency. The PRAG is charged to perform the following duties:

- Ensure that the *Peer Review Handbook* is updated periodically.

- Develop peer review training for the agency.

- Provide expert advice to the STPC regarding peer review issues.

- Develop products for internal and external release that advance peer review in the Agency.

- Serve as a forum for discussing issues or questions relating to peer review.

2.2.4. What Is the Role of the Office of the Science Advisor?

OSA, with assistance and cooperation from all EPA program and regional offices, is responsible for producing the Agency's annual report to Office of Management and Budget (OMB) that summarizes the peer reviews that were conducted during the previous fiscal year for Influential Scientific Information (ISI), including Highly Influential Scientific Assessments (HISAs). OSA also provides support to the STPC and PRAG on peer review activities.

2.2.5. What Is the Role of the Office of Research and Development?

ORD is responsible for maintaining the EPA Science Inventory (SI) database. In addition, ORD maintains the EPA Peer Review Agenda website[11] that meets the OMB Peer Review Bulletin guidelines for a publicly available, "web-accessible listing of forthcoming influential scientific disseminations ... that is regularly updated by the agency" (see Appendix B). For information on the SI and Peer Review Agenda, see Section 7.3.

2.3. Peer Review Roles and Responsibilities within EPA Offices

EPA program and regional offices are responsible for carrying out all aspects of peer review appropriate for their work products. This includes categorizing their work products as ISI, HISAs or "other," as well as determining the nature, scope and timing of the peer review and following the procedures outlined in this Handbook. For ensuring greater independence and transparency of peer reviews, it is important to separate the responsibilities for developing work products from conducting the peer review (see Figure 2), whenever possible. The roles of individuals with specific responsibilities for peer review within their organization are addressed in the following subsections.

---

[11] EPA. *Peer Review Agenda.* http://cfpub.epa.gov/si/si_public_pr_agenda.cfm.

2.3.1. What Is the Role of the Assistant and Regional Administrators?

The EPA's AAs and RAs are responsible for all peer review actions in their organizations. In many cases, the AA or RA may delegate these responsibilities to a DM (e.g., DAA, DRA, and Office/Division Director) within their organization. When more than one EPA office or other agencies are involved in the development of a work product, responsibility for conducting the peer review can be negotiated; often, the degree of involvement by any of the organizations and agencies and their ability to fund peer review will determine who assumes the lead for the peer review.

As part of the annual review process, AAs and RAs ensure that the peer review of influential scientific and technical work products in their program or regional office has been conducted and documented appropriately.

2.3.2. What Is the Role of the Decision Maker?

The DM should ensure that there are processes in place to determine—early in the planning stage of the product—whether the product is (or is likely to be) influential, and if influential, whether it is (or is likely to be) a HISA, and determine how the peer review is to be conducted. As noted in Section 2.3.1, the AA/RA may delegate these responsibilities to a manager within the organization, such as the ORD Laboratory or Center Director, Program Office Director, or Regional Division Director.

Specific responsibilities of the DM are the following:

- Determine which type of work products need to be peer reviewed and the nature of the peer review to be conducted for each type, and ensuring compliance with all applicable guidance (including the OMB Peer Review Bulletin).

- Identify the stages of product development for which peer review is appropriate and decide how the peer review is to be conducted.

- Document the categorization determination and other peer review planning decisions (see Roadmap Exhibit 3, Example EPA Peer Review Decision Summary Documentation), especially if the product is (or is likely to be) influential, and if influential, whether it is (or is likely to be) a HISA.

- Designate a Peer Review Coordinator (PRC) within the organization.

- Designate a Peer Review Leader (PRL) to plan, conduct and complete the peer review. The person in charge of producing the work product (Principal Investigator, Project Leader, or Project Manager (PM) – see Section 2.4.4) may serve as the PRL; however, for ISI and HISAs, the DM should consider the advantage of designating a different individual to serve as the PRL to enhance the independence of the peer review process.

- Ensure that sufficient funds are designated in the EPA office's budget to conduct the peer review and allocate adequate resources throughout the peer review process (e.g., contractor support for peer review).

- For HISAs, decide whether it is feasible and appropriate to make the draft scientific assessment available to the public for comment before or at the same time it is submitted for peer review, and whether it is feasible and appropriate to sponsor a public meeting at which oral presentations on scientific issues can be made to the peer reviewers by interested members of the public.

- Ensure that all relevant issues and comments raised by the peer reviewer(s) are adequately addressed and documented for the record and, when appropriate, incorporated into the final work product.

2.3.3. What Is the Role of the Peer Review Coordinator?

The PRC is designated by the DM to coordinate and monitor all peer review activities related to EPA scientific and technical work products in an organization. This individual has access to senior management and all staff across the organization involved with peer review, and is the main contact with the PRAG, OSA and ORD for information about peer review activities and submissions to the SI.

Although some of the following functions might be performed by other personnel, specific responsibilities of the PRC are the following:

- Work closely with the DM and PRL to plan the peer review of the work product and ensure that peer review guidelines and procedures are appropriately applied.

- Provide advice, guidance and support to the PRL and, as determined by management, serve as the PRL for certain work products.

- Establish procedures to ensure that the peer review process is adequately documented in a peer review record (see Section 6.5) and that the record is filed and maintained in a manner consistent with Agency retention policies.

- For ISI and HISAs, ensure that information in the peer review record is consistent with OMB reporting guidelines by making key pieces publicly available on the Agency's Peer Review Agenda[12] via the SI.

- Deliver peer review training to management and staff.

- Function as the liaison with the PRAG, OSA and ORD by participating in PRAG workgroups as needed.

- Ensure that the list of work products and their associated peer review mechanisms are accurate and updated during the annual reporting (and, when necessary, at other times).

- Post or link other relevant peer review documents to the PRA from the SI.

2.3.4. What Is the Role of the Peer Review Leader for EPA-Managed Peer Reviews?

The PRL plans, conducts and completes the peer review for specific work products within an organization. The PRL is selected by the DM. To enhance the independence of the peer review process, the DM should consider the advantage of having separate individuals produce the work product and manage the peer review (see Section 2.3.2). The PRL should follow the Agency's peer review

---

[12] EPA. *Peer Review Agenda*. http://cfpub.epa.gov/si/si_public_pr_agenda.cfm.

procedures and guidelines and should receive training on the Handbook and other policies and guidelines applicable to peer review. For peer reviews conducted by outside organizations such as the National Academy of Sciences (NAS), the PRL should be thoroughly familiar with the ethics policies and requirements of the organization conducting the review (see Section 5.3.1).

Specific responsibilities of the PRL include:

- **Plan the peer review:** After considering the type of work product under development, the PRL (in consultation with the DM and PRC) should do the following:

  o Determine and document the categorization of the product (ISI, HISA or other) and when and how the peer review should occur.

  o Establish a plan for the peer review, including the peer review approach (e.g., letter, panel, journal, EPA- or contractor-managed peer review); the scope and timing of the peer review; and the approach to responding to peer review comments.

  o Obtain management approval of the plan, and ensure proper documentation of decisions as part of the peer review record.

  o Develop the charge for the peer reviewers, soliciting input from the project team developing the work product and the public, as appropriate. When the timing of panel selection does not allow for prior finalization of the charge, develop a preliminary version of the charge that provides enough detail about anticipated peer review scope and issue areas that requisite areas of peer review panel expertise can be identified.

  o Select peer reviewers with expertise appropriate for the charge after considering and resolving any ethics issues, including potential conflicts of interest (COIs).

  o Ensure that appropriate internal review, including clearance procedures, is completed before releasing the product for external peer review.

- **Conduct the peer review:** The PRL should:

  o Provide opportunities for public comment on the review materials, when applicable (usually for ISI or a HISA).

  o Provide the peer reviewers with materials relevant to the work product, including instructions; the charge questions; and significant scientific and technical comments, if public comment was sought. Particularly for HISA, include information about key studies or models used to support key findings or conclusions of the work product.

  o Advise peer reviewers of their responsibility to prepare their response to the charge, usually in the form of a report documenting the results of the peer review.

  o Document any changes to the charge, profile of peer reviewers or ethical conflicts that may develop, and keep the PRC informed throughout the process.

- **Complete the Peer Review:** To complete the peer review, the PRL should:

  ○ Ensure that peer review comments are incorporated, as appropriate, into the final work product.

  ○ Document the resolution in a "response to comments" or a "reconciliation memorandum," clearly identifying comments that have not been addressed.

  ○ Obtain the DM's approval on the resolution of peer review comments.

  ○ For ISI and HISAs, make the peer review report (see Table 1) and any Agency response to comments publicly available on the Agency's Peer Review Agenda.[13]

  ○ For ISI and HISAs, inform the PRC when the peer review is completed and available for inclusion in the annual report to OMB (see note in Section 6.4).

  ○ Archive the peer review record in a manner consistent with the organization's records management procedures.

2.3.5. What Are the Roles of the Peer Review Leader and Contractor in the Case of Contractor-Managed Peer Reviews?

Several responsibilities of the PRL will shift to a contractor when a contractor is managing the peer review, but the PRL still ensures the peer review is conducted and completed for a specific work product following Agency procedures. For example, consistent with the contract terms, the contractor is responsible for selecting peer reviewers with due consideration of ethics issues (such as potential COIs or an appearance of a loss of impartiality [see Section 4.6]) and the balance of expertise, providing review materials and instructions to the peer reviewers and compiling the peer reviewer comments. The PRL provides materials associated with the peer review to the Contracting Officer's Representative (COR), who is the technical point of contact for the contract. In some cases, the PRL and the COR may be the same individual. The COR then provides the materials to the contractor, who distributes them to the peer reviewers. After the peer review, the contractor ensures that the reviewers have fulfilled their responsibilities under their agreement with the contractor. EPA should not alter the contractor's peer review report. The contractor may have additional responsibilities, depending on the complexity of the peer review and public participation in the process. For more information on contractor-managed peer reviews, see Section 4.6.

2.3.6. What Is the Role of the Designated Federal Officer (DFO) in the Case of Federal Advisory Committee (FAC)-Conducted Peer Reviews?

When peer reviews are conducted through a FAC, some of the PRL responsibilities are assumed by the DFO. The DFO is an EPA employee who is responsible for managing the FAC and ensuring that the provisions of the Federal Advisory Committee Act (FACA) are met (see Section 4.7). Details of the duties and responsibilities of DFOs are available in the Agency's *Federal Advisory Committee Handbook*.[14] For example, when external peer review is conducted under the auspices of the Science Advisory Board (SAB) or the Clean Air Scientific Advisory Committee (CASAC), the SAB Staff Office

---

[13] EPA. 2015. *Peer Review Agenda*. http://cfpub.epa.gov/si/si_public_pr_agenda.cfm.

[14] EPA. 2013. *Federal Advisory Committee Handbook*. BiblioGov.

in the Office of the Administrator is responsible for selecting and vetting independent experts; planning, budgeting for and conducting peer review meetings; and maintaining peer review committee records.

The SAB Staff Office selects peer reviewers after a public nomination and comment process and after evaluating candidates for potential COIs or appearance of a loss of impartiality. The SAB Staff Office also announces committee meetings in the *Federal Register* and on the committee website, prepares detailed meeting minutes, transmits EPA charge and review materials to the committee and provides support to the committee in preparation of the advisory report to the EPA Administrator. To maintain the independence of the peer review process, the SAB Staff Office does not draft the EPA charge or prepare the Agency response to the peer review. The SAB Staff Office also does not enter data into the SI.

2.3.7. What Are the Roles and Responsibilities of EPA When Peer Reviews Are Conducted by the National Academy of Sciences?

The NAS is a private, nonprofit society of distinguished scientists established by Congress to provide independent, objective advice to the nation on science and technology matters. When agencies request an NAS peer review or sponsor an NAS study, a contract mechanism is used. The Agency works with NAS staff to develop a set of charge questions called a "statement of task" and also helps to define the timing and cost of the review. NAS reviews usually are conducted through the National Research Council (NRC). Once the statement of task and budget are approved by the NRC Governing Board, responsibilities for the peer review and products lie with the NAS and not EPA. The EPA contact with the NAS is a COR, and there can be more than one COR associated with an EPA-sponsored NAS review.

2.3.8. What Are the Roles and Responsibilities of EPA Authors and Managers Associated With Journal Peer Review?

The EPA considers peer review by a refereed scientific journal to be a satisfactory form of peer review to determine the scientific credibility and validity of the scientific and technical information presented in the article. Because journal peer review is an example of external review, the DM and PRL (typically one of the authors) have responsibilities for this type of peer review. The EPA authors of the article are responsible for complying with relevant organizational procedures associated with publications, such as internal review and clearance prior to submission to a journal; complying with pre-dissemination requirements, such as the use of an appropriate disclaimer; addressing peer review comments and responding to the editor; and maintaining a record of the peer review process. Peer-reviewed journal articles should be submitted to the SI as appropriate.

2.4. Other Agency Personnel Involved With Peer Review

2.4.1. What Are the Roles of the Offices of General and Regional Counsel?

OGC and ORC attorneys have specific areas of expertise, such as contracts and procurement, ethics and the FACA. They are consulted as needed to assist EPA staff with their oversight responsibilities. OGC/ORC attorney review and involvement helps ensure that Agency peer reviews meet legal standards, including those for integrity, transparency and openness.

2.4.2. What Are the Roles of the Quality Assurance Manager (QAM), Director of Quality Assurance (DQA) and Quality Assurance (QA) Staff?

The QAM, DQA and QA staff oversee implementation of the organization's Quality System pursuant to the EPA's Quality Policy for environmental data collection and use (see Section 1.3.4). QA processes and procedures are essential for developing scientifically sound, transparent and credible information supporting EPA's products and decisions. Typically, the QA staff conducts technical review of data quality and review of scientific and technical products for consistency, correctness, coherence, clarity and conformance. In planning the peer review, the PRL is encouraged to consult with the organization QA contact to determine documentation of QA requirements. If applicable, the PRL should ask the QAM to review the QA statement or QA section included in the draft or final work product.

2.4.3. What Is the Role of the Information Quality Guidelines (IQG) Officer?

The IQG Officer (or Coordinator) assists the organization in establishing pre-dissemination review procedures for the quality, objectivity, utility and integrity of the EPA's information products disseminated to the public. The PRL, PRC, QAM and DQA can collaborate with the IQG Officer to ensure compliance with the organization's established pre-dissemination procedures for the specific work products disseminated by EPA.

2.4.4. What Is the Role of the Principal Investigator (PI), Project Leader (PL) or Project Manager (PM)?

The PI, PL or PM is responsible for producing work products based on sound scientific principles and practices, and is responsible for working with the PRL to get their work products peer reviewed. The Agency's peer review procedures and guidelines, Quality Policy requirements for use of defensible data, the General Assessment Factors guidance and the Scientific Integrity Policy provide the framework for assuring the integrity and utility of the EPA's science. The PIs, PLs and PMs are expected to be familiar with these policies. The PI, PL and PM should work collaboratively with the PRC and PRL throughout the peer review process and should help develop charge questions specific to the work product. To enhance the independence of the peer review process for ISI/HISAs, a separate PRL, rather than the PI, PL or PM, should be considered to manage the peer review.

2.4.5. What Is the Role of the Contracting Officer's Representative (COR)?

For some peer reviews, a contractor takes on some of the roles of the PRL. The Contracting Officer (CO) can delegate some responsibilities to the COR. The COR is sometimes called the Project Officer, Task Order Project Officer or Work Assignment Manager. The COR provides oversight of the peer review process. In some instances, the PI, PL or PM can serve as the COR. When a contractor-managed peer review approach is used, the PRL works with and through the COR for some activities. The COR, together with the CO, is responsible for ensuring compliance with contracting requirements, developing a Statement of Work (SOW), coordinating with the contractor regarding COI and other administrative matters and overseeing contractor activities to ensure that the schedule and other contract requirements are met. Unless they also are the COR, the PI, PL or PM cannot supply materials directly to the contractor. Responsibilities of the CO also are described in Section 4.6, especially as they relate to the inclusion of COI solicitation provisions and contract clauses. In accordance with the EPA's peer review process for contractor-managed panels of ISI and HISAs, when consultation about COI is needed between the EPA Science Advisor and contractors, the CO and COR should participate in the consultation.

In some cases, the Agency may opt to obtain peer review services directly from individual peer reviewers, rather than through a contractor-managed peer review process. In such cases, the Agency generally would use a Purchase Order to compensate external peer reviewers, and the Agency contact would be the Purchasing Agent or the COR, if one is designated.

3. Categorize the Work Product and Determining the Need for Peer Review

3.1. Overview

The EPA produces or uses a variety of scientific and technical work products. Before a peer review approach can be selected, a determination first must be made and documented about whether the scientific or technical work product is influential scientific information (ISI) as defined by the Office of Management and Budget's (OMB) Peer Review Bulletin.[15] Although other scientific work products may benefit from peer review, peer review should be conducted for those that are categorized as influential. Influential scientific and technical work products generally receive internal peer review, followed by external peer review. Other work products that do not meet the OMB definition of influential products may undergo internal peer review, external peer review or both.

This chapter of the Handbook describes products that might be subject to peer review, how EPA determines whether a scientific and technical work product is influential—including whether it is a Highly Influential Scientific Assessment (HISA), which is a subset of ISI—and the critical role of senior managers in that decision (Figure 4). The distinction between ISI and HISAs is important because there are additional peer review considerations for HISAs.

3.1.1. What Are Scientific and Technical Work Products?

The first step in determining which work products should be peer reviewed is to identify those that are scientific or technical in nature. The term "scientific and technical work products" is generally consistent with the term "scientific information" in the OMB Peer Review Bulletin. Scientific and technical work products are used to support a research agenda,

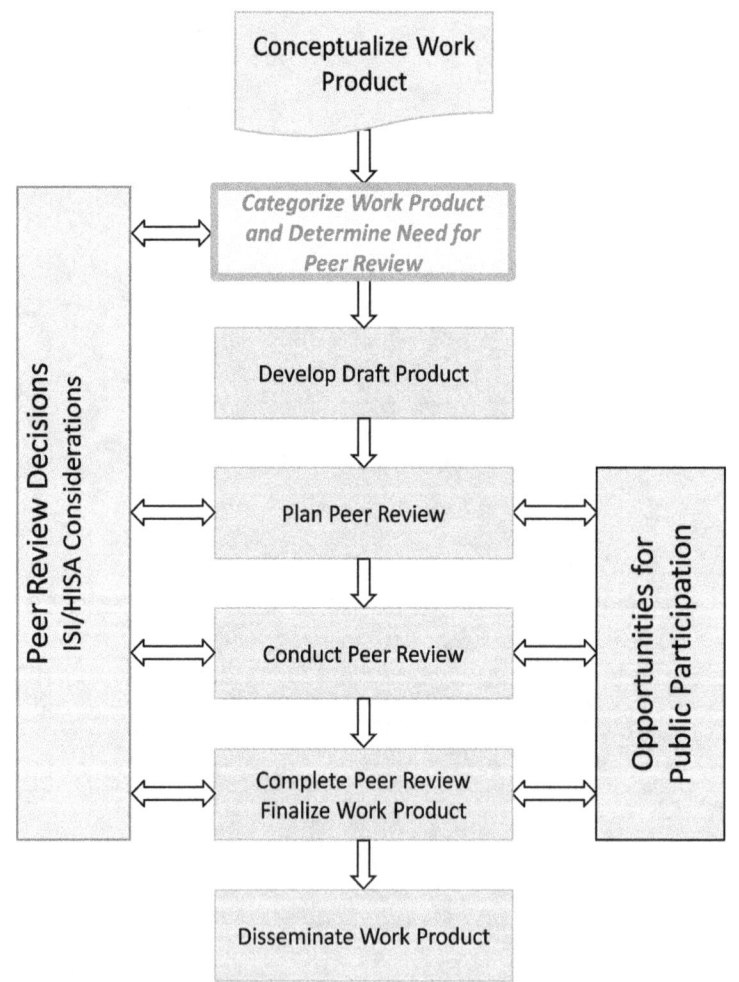

Figure 4. The Peer Review Process: Develop and Categorize Work Product/ Plan Peer Review

---

[15] OMB defines "scientific information" as "factual inputs, data, models, analyses, technical information, or scientific assessments based on the behavioral and social sciences, public health and medical sciences, life and earth sciences, engineering, or physical sciences." (OMB Peer Review Bulletin, Section I.5).

regulatory program, policy position, or other EPA position or action. Scientific and technical work products include economic and social science work products. Categories of work products include, for example, risk assessments, technical studies and guidance, analytical methods, scientific database designs, technical models, technical protocols, statistical surveys/studies, technical background materials, technical guidance (except for guidance providing policy decisions), research plans and research strategies.

Products that would <u>not</u> be considered scientific or technical work products can include the following:

- Products that address procedural matters (e.g., planning, reporting, coordination, notification).

- Primarily policy statements (e.g., relocation policy).

- Conference proceedings (unless the proceedings are used as the scientific basis for an Agency action or decision).

- Decision documents, such as an Environmental Impact Statement (EIS), Record of Decision (ROD), or an Economic Analysis reviewed through an interagency review process under E.O. 12866.

- Products that summarize a scientific and technical work product, including public affairs and communication materials (e.g., press releases, press kits, brochures, fact sheets); scientific abstracts, including posters and presentations at scientific meetings; or other summaries (e.g., summaries on Web pages).

- Strategic plans, Agency annual plans and budget documents, performance reports, analytical blueprints, and goals documents.

For any of these examples, the document itself is not subject to the Peer Review Policy, but the underlying scientific or technical models, data and/or work products upon which these documents are based are candidates for peer review. Scientific and technical work products that are referenced to provide context, history, or general background information and that do not materially influence or educe an agency policy or action generally need not undergo peer review.

3.1.2. Who Develops Scientific and Technical Work Products?

Scientific and technical work products may be generated by one or more EPA offices or in collaboration with external partners.[16] Scientific and technical products also may be generated by third-party organizations and used by EPA. In general, third-party scientific and technical products should be evaluated for peer review if they will be used to support Agency decisions or actions.

---

[16] Please note that generation of scientific or technical work products in collaboration with external partners may be subject to the Federal Advisory Committee Act (FACA).

3.1.3. What Scientific and Technical Work Products Need Peer Review?

According to the EPA's Peer Review Policy, "[p]eer review of all scientific and technical information that is intended to inform or support agency decisions is encouraged and expected." The OMB Peer Review Bulletin stipulates that all of the agency's ISI and HISAs should be peer reviewed unless they meet exemption criteria (see Section 3.3). Other scientific work products that do not rise to the level of influential also may be peer reviewed. These work products will have greater standing in the scientific community if an independent peer review is completed.

> *When in doubt about whether a work product merits peer review, decide to peer review it.*

New applications or modifications of existing, adequately peer-reviewed methodologies or models that significantly depart from the situations for which they were originally designed may require additional peer review.

3.2. Assignment of Categories

3.2.1. What Is Influential Scientific Information (ISI)?

As defined by the OMB Peer Review Bulletin, the term "influential scientific information" means scientific information the agency reasonably can determine will have or does have a clear and substantial impact on important public policies or private-sector decisions. The interpretation of the term "influential" is consistent with OMB's government-wide information quality guidelines (IQG)[17] and the IQG of the Agency. (The Agency has linked its use of the term "influential" to the term "major" in its IQG).

At EPA, scientific and technical work products that will have or do have a clear and substantial impact on important public policies or private-sector decisions would be considered influential. Decision Makers (DMs) should consider the following factors when determining whether a product is likely to be influential:

- Establishes a significant precedent, model or methodology.

- Is likely to have an annual effect on the economy of $100 million or more, or adversely affect in a material way the economy; a sector of the economy; productivity; competition; jobs; the environment; public health or safety; or state, tribal or local governments or communities.

- Addresses significant controversial issues.

- Focuses on significant emerging issues.

- Has significant cross-agency and/or interagency implications.

- Involves a significant investment of agency resources.

---

[17] OMB. 2002. *Guidelines for Ensuring and Maximizing the Quality, Objectivity, Utility, and Integrity of Information Disseminated by Federal Agencies; Republication. Federal Register* 6: 8,452. February 22.

- Considers an innovative approach for a previously defined problem, process, or methodology.

- Satisfies a statutory or other legal mandate for peer review.

3.2.2. How Are ISI Determinations Made and Documented?

The DM, in consultation with the Peer Review Leader (PRL), should make the judgment as to whether a work product is ISI and document the decision. Generally, determination of whether a scientific and technical work product is influential will occur on a case-by-case basis. The EPA's work products should be evaluated and assessed with respect to the factors defined in Section 3.2.1. The categorization determination and other peer review planning decisions should be documented (see Roadmap Exhibit 3: *Example EPA Peer Review Decision Summary Documentation*).

3.2.3. What Is a Highly Influential Scientific Assessment (HISA)?

HISAs are a subset of ISI for which the OMB Peer Review Bulletin specifies additional peer review considerations, including that peer reviewers be external, non-EPA experts. OMB has defined a HISA as ISI that "the agency or the Administrator determines to be a scientific assessment that:

    (i) could have a potential impact of more than $500 million in any year, or

    (ii) is novel, controversial, or precedent-setting or has significant interagency interest."

OMB defines a scientific assessment as "an evaluation of a body of scientific or technical knowledge, which typically synthesizes multiple factual inputs, data, models, assumptions, and/or applies best professional judgment to bridge uncertainties in the available information."[18] Examples given by OMB of assessments that may be considered HISAs include: state-of-science reports; technology assessments; weight-of-evidence analyses; meta-analyses; health, safety or ecological risk assessments;[19] toxicological characterizations of substances; integrated assessment models; hazard determinations; or exposure assessments.

The more far-reaching or significant the impacts of a scientific assessment, the more appropriate it is to categorize the product as a HISA. If a work product is a scientific assessment that involves significant issues that truly are "cutting-edge," it might be appropriate to designate it as a HISA. For examples of HISA products, see the Science Inventory or the Peer Review Agenda (http://cfpub.epa.gov/si/si_public_pr_agenda.cfm).

3.2.4. How Are HISA Determinations Made and Documented?

Once a scientific or technical assessment has been determined to be influential, the DM should determine whether the product meets OMB's definition of a HISA. As with the categorization of a work product as influential, the decision whether or not to elevate a scientific assessment to the highly influential category occurs on a case-by-case basis after considering the criteria discussed in Section 3.2.3. The DM should make the judgment as to whether an assessment is a HISA and the

---

[18] OMB Peer Review Bulletin, Section I.7.

[19] Influential scientific information regarding human health, safety or environmental risk assessments may be subject to quality principles articulated in Section 6.4 of the *Guidelines for Ensuring and Maximizing the Quality, Objectivity, Utility, and Integrity of Information Disseminated by the Environmental Protection Agency* (2002, EPA/260R-02-008).

decision should be documented (see Roadmap Exhibit 3, Example EPA Peer Review Decision Summary Documentation).

3.2.5. What Work Products Are Categorized as "Other"?

Any scientific and technical work product that does not meet the OMB guidelines' criteria for influential information is categorized as an "other" work product. Examples may include, but are not limited to, journal articles and some reports. The OMB Peer Review Bulletin does not apply to journal articles because such publications do not contain findings or conclusions that represent the official position of the Agency.

3.2.6. Are Work Products Categorized as "Other" Candidates for Peer Review?

Yes, the Agency may decide to use peer review for work products categorized as "other" because of a particular EPA office's needs and goals. Peer review also may be warranted because it adds substantial value to the work product or if the work product will be used in an Agency decision-making process. Research papers submitted to peer-reviewed scientific journals are categorized as "other" yet still undergo peer review by the journal.

3.2.7. Can the Categorization of a Work Product Be Revised After the Peer Review Planning Phase?

Yes, the categorization can be revised after the peer review planning phase but before the product undergoes peer review. The nature of the work product—or its intended use—may change, so re-evaluation may be necessary to ensure an appropriate peer review is conducted.

Furthermore, the impact and interest in a peer-reviewed scientific product may change or may not be anticipated fully by the PRL or the DM. Under such circumstances, additional peer review may be necessary, including a change in the review mechanism. Any decision to modify the categorization of a work product should be documented in the peer review record (see Section 6.5.2).

3.3. Influential Work Products That Are Not Peer Reviewed

3.3.1. Under What Circumstances Are Influential Work Products Exempt From the Provisions of the OMB Peer Review Bulletin?

Per the OMB Peer Review Bulletin, the following information does not need to be peer reviewed, even if it might be considered ISI or a HISA:

- Information related to certain national security, foreign affairs or negotiations involving international trade or treaties for which peer review would interfere with the need for secrecy or promptness.

- Information disseminated in the course of an individual adjudication or permit proceeding (including a registration, approval, licensing or site-specific determination), unless the Agency determines that peer review is practical and appropriate and the influential information is scientifically or technically novel or likely to have precedent-setting influence on future adjudications and/or permit proceedings.

- Information involving a health or safety issue where the Agency determines that the dissemination is time-sensitive.

- A regulatory impact analysis or regulatory flexibility analysis subject to interagency review under Executive Order 12866, *Regulatory Planning and Review*,[20] except for underlying data and analytical models used.

- Routine statistical information (e.g., periodic demographic and economic statistics) and analyses of these data to compute standard indicators and trends.

- Accounting, budget, actuarial and financial information.

- Information disseminated in connection with routine rules that materially alter entitlements, grants, user fees or loan programs, or the rights and obligations of recipients thereof.

3.3.2. Are There Other Circumstances When Peer Review of Influential Products Is Not Necessary?

Yes, there are other circumstances when peer review of influential products may not be necessary. For example, peer review generally is not conducted:

- For work that has been reviewed previously in a manner consistent with the OMB Peer Review Bulletin and this Handbook (e.g., a cancer risk assessment methodology or an exposure modeling technique that was the subject of earlier peer review of appropriate technical merit would not generally undergo additional peer review even if the product supported a significant Agency decision).

- If an application of an adequately peer-reviewed work product does not depart significantly from its scientific or technical approach.

- When the scientific or technical methodologies or information being used are commonly accepted in the field of expertise and have the appropriate documentation to support the commonly held view (e.g., many products supporting Control Techniques Guidelines and Effluent Limitation Guidelines).

- When the product was developed by the National Academy of Sciences (NAS).

3.3.3. For Influential Information That Is Not Exempt, Can the Peer Review Provisions of the OMB Peer Review Bulletin Be Waived or Deferred?

The Administrator may waive or defer the peer review provisions of the OMB Peer Review Bulletin for ISI (including IIISAs) if there is a compelling rationale for the waiver or deferral. The use of waivers is expected to be limited to unusual and compelling situations not otherwise covered by the exemptions, such as situations in which unavoidable legal deadlines prevent full implementation of the OMB Peer Review Bulletin's peer review provisions. According to the Bulletin, deadlines found in consent decrees ordinarily will not warrant waiver of the provisions because those deadlines should be negotiated to

---

[20] Executive Order No. 12866. October 4, 1993. *Federal Register,* 51:735. http://www.archives.gov/federal-register/executive-orders/pdf/12866.pdf.

permit time for conducting a peer review. Deferral of some or all of the peer review provisions may be an appropriate way to accommodate immovable deadlines. If any of the OMB Peer Review Bulletin provisions are deferred, peer review should be conducted as soon as practicable thereafter. Deferrals of peer review of ISI and HISAs should be approved by the Administrator.

If peer review of an influential work product is not planned, an explanation should be included in the product documentation and record for that work product in the Science Inventory (SI).

3.4. Work Products from Contracts, Grants and Agreements That May Require Peer Review

The Agency should not use scientific and technical work products from contracts, grants or cooperative agreements to support decision making unless the work products have undergone a peer review both for scientific and technical rigor and for applicability to the specific use to be made of the product. Products generated by contractors under the direct supervision of EPA and incorporated by the Agency in the development of EPA scientific and technical work products are not necessarily peer reviewed separately but as part of the final Agency product.

Contracts differ from grants and cooperative agreements and require special considerations when considering peer review of these work products (see Section 3.4.2). There are important legal restrictions on the direct use of work products developed under grants and cooperative agreements in the agency's decision-making process. See the EPA's Grants and Debarment Web page (http://www.epa.gov/ogd/ or http://intranet.epa.gov/OGD/policy/7.0-GPI-GPI-94-04.htm) for additional information.

3.4.1. How Does the EPA's Peer Review Process Apply to Products Generated through EPA Contracts?

A work product generated through an EPA contract should undergo the same degree of peer review as if the work product was developed by an EPA employee. The peer review should be conducted independently from the contractor who developed the work product. EPA is responsible for arranging the peer review (see Section 4.6.1).

3.4.2. How Does the EPA's Peer Review Process Apply to Products Generated through EPA Assistance Agreements (e.g., Grants or Cooperative Agreements)?

Special considerations apply to the peer review of scientific and technical work products generated through EPA grants or cooperative agreements.

EPA provides financial assistance for research that is intended to stimulate or support development of scientific knowledge that is not primarily for EPA's direct use or benefit. The resulting work products might be widely disseminated either through publication in scientific journals or through other means, as opposed to a report tailored to the EPA's specific needs and requirements. EPA can consider these work products just as it does other published scientific works when formulating its programs and policies. EPA may determine that the recipient's work product is influential because (1) it will be used to support an EPA program or policy position; and (2) it meets the criteria for influential information. EPA should evaluate whether the peer review process undertaken by the assistance agreement recipient was acceptable for the purposes for which EPA plans to use the work product. EPA may accept the peer review if it determines that it is of appropriate quality and as defensible as if it were conducted by EPA itself. The work product may require additional peer review, however, in the context of its use or modification by the Agency.

The following are options for peer reviewing the product:

- EPA can have the product peer reviewed with the participation of the assistance agreement recipient/author(s). In this case, EPA could arrange for an independent peer review of the product within the context of the way(s) in which the Agency plans to use it. EPA may ask the recipient/author(s) to provide additional information or to revise the product in response to the peer review.

- EPA can have the product peer reviewed without the participation of the recipient/author. EPA could arrange for the peer review of the product within the context of the Agency's intended use. EPA then would receive the comments and prepare a statement that documents the EPA's own response to the comments.

3.4.3. Can the Recipient of a Grant or Cooperative Agreement Use Agreement Funds to Pay Peer Reviewers of Their Work Products?

Provided that EPA agrees that a peer review would further the public purpose of the assistance agreement, EPA may include funds for the peer review in the agreement. This is generally in the form of journal publication fees. If a work product is ISI or a HISA, the peer review of that product should follow the guidelines set out in the *Peer Review Handbook*, consistent with Agency use and review of the product.

3.4.4. How Should Peer Review Be Handled for Products Developed Under an Interagency Agreement?

Under an Interagency Agreement, EPA provides funds to another agency to be used for a specific purpose. The receiving agency's guidance for peer review is likely to be different from the EPA's Peer Review Policy, although the OMB Peer Review Bulletin establishes some minimum common guidance for the federal government. Regardless, if EPA plans to use any work products from that agreement, a determination should be made as to whether the work products are ISI, including whether they are HISAs, or do not qualify as influential (i.e., "other"). The EPA then should decide whether those documents need review under the EPA's Peer Review Policy and pursue the appropriate mechanism.

3.5. Other Types of Work Products That May Require Peer Review

3.5.1. Should Another Organization's Work Products That Have Been Submitted to the EPA for Use in Decision Making Be Peer Reviewed?

Any scientific or technical work product that is used in agency decision making and is considered influential becomes a candidate for peer review, regardless of whether the work product is developed by EPA or another organization. Therefore, all work products important to EPA decision making that are independently generated by other organizations (e.g., other federal agencies, interagency groups, state and tribal bodies, environmental groups, industry, educational institutions, international bodies) should be considered as candidates for peer review. The DM in the EPA office planning to use the product is responsible for the categorization and decision regarding peer review.

If possible, when EPA knows that a work product being generated by another organization may be of interest to EPA for future use, the appropriate EPA office(s) should work with that organization and others, as appropriate (e.g., state agencies, international organizations), to promote the use of peer review. Furthermore, when another agency's product is being considered for EPA use, the EPA office(s) planning to use the product should ascertain—in collaboration with other EPA offices as appropriate—the characteristics and sufficiency of any peer review process already conducted or planned for the candidate product.

Reports produced by certain outside organizations—such as the NAS, the EPA's Science Advisory Board (SAB) and the International Agency for Research on Cancer—are products of independent peer review by their nature. The OMB Peer Review Bulletin specifically notes that official NAS reports are generally presumed not to require additional peer review. The Agency's scientific work products which use and interpret those products' findings or results may be subject to peer review. Peer reviews conducted by stakeholders of their own products may be considered peer input but not independent peer review, unless principles and policies articulated in the EPA's *Peer Review Handbook* can be applied.

3.5.2. Is Additional Peer Review Necessary If a Paper Is Published in a Refereed Scientific Journal?

The extent to which additional peer review is needed for an article that has been peer reviewed by a credible refereed scientific journal depends upon EPA's use of the article. For example, EPA may determine that an additional and more rigorous or transparent review process is needed if a particular journal review process did not address questions that EPA determines should be addressed before using or disseminating the information.

3.5.3. Does an Agency Work Product Become a Candidate for Peer Review When Peer-Reviewed Journal Articles Are Used in Support of That Work Product?

Agency work products are candidates for peer review even when supported by peer-reviewed journal article(s). Although the use of articles that have been peer reviewed by a credible journal strengthens the scientific and technical credibility of any work product in which the article(s) appears or is referenced, it does not eliminate the need to consider whether the work product itself should be peer reviewed. In most cases, journal peer review may not cover issues and concerns that the Agency may want peer reviewed to support an EPA action. Under these circumstances, the scientific or technical work product in which the article(s) appears or is referenced becomes a candidate for peer review. A journal article authored by EPA employees should be used in the same manner as an article published by non-EPA authors in a credible, well-recognized journal.

Decisions to peer review a work product should be documented in the peer review record (see Section 6.5.2).

3.5.4. Should Site-Specific Decisions Be Subject to Peer Review?

A site-specific decision (e.g., for a permit or hazardous waste cleanup) itself is not subject to peer review under the EPA's Peer Review Policy. However, if a site-specific decision is supported by ISI or a HISA generated for that site-specific decision, then that work product should be peer reviewed. Generally speaking, the PRL should examine closely the ways in which the underlying scientific or technical work product is adapted to the site-specific circumstances.

3.5.5. Should National Environmental Policy Act (NEPA) Products Be Subject to Peer Review?

Although an EIS prepared under the requirements of the NEPA receives extensive review through the "scoping" and interagency and public review processes that are part of the NEPA, this usually is not considered peer review. If the underlying scientific or technical data, models, analyses or work products are categorized as ISI or a HISA, then these should be peer reviewed.

If EPA is developing the NEPA document as part of an EPA action/decision (i.e., EPA is the lead agency under NEPA), and supporting documents are ISI or HISAs, then the supporting documents should receive independent peer review. If the document is not categorized as influential, then peer input might be appropriate.

If EPA is reviewing an EIS from another agency (i.e., EPA is not the lead agency under NEPA), it is likely that it is being reviewed for conflicts with EPA policy and general environmental concerns. In such a case, EPA should ask whether the underlying scientific or technical work product that supports the EIS has been peer reviewed to avoid concerns about the full credibility and soundness of the EIS based on the science and technical support. The EPA should work with the other organization/agency to ensure that scientific and technical work products receive peer review adequate for EPA purposes.

3.5.6. Do Voluntary Consensus Standards Undergo Peer Review?

In general, the answer is no. The National Technology Transfer and Advancement Act of 1995 (NTTAA) directs EPA to use available voluntary consensus standards in its regulatory activities, unless to do so would be inconsistent with applicable laws or otherwise impractical. For purposes of the NTTAA, voluntary consensus standards are defined as technical standards (e.g., materials specifications, test methods, sampling procedures, business practices) that are developed or adopted by voluntary consensus bodies (e.g., ASTM International). The general purpose of the NTTAA is to reduce private and governmental costs by avoiding having the government "reinvent the wheel" in the development of technical standards. Voluntary consensus standards normally would not undergo peer review because the underlying process used by issuing organizations to develop and approve these standards generally is considered adequate for purposes of the Agency's Peer Review Policy.

3.5.7. What Economic Work Products Need Peer Review?

Economic work products are considered scientific and technical work products. As such, it may be appropriate to peer review them, and an ISI/HISA/other determination should be made. If an economic work product is determined to be influential, then it should be peer reviewed if it has not been subjected already to adequate peer review according to the relevant sections of this Handbook or is otherwise exempt (see Section 3.3).

Data and analytical models underlying an economic analysis, particularly those supporting economically significant rules, are candidates for peer review if the models and corresponding use of the data have not been subjected previously to adequate peer review. This also is true for work products that will serve as a principal method or protocol used to conduct economic analyses within a program.

The following economic work products generally should be peer reviewed:

- Internal Agency guidance for conducting economic and financial analysis that meets the definition of influential.

- Economic and financial methodologies that will serve as a principal method or protocol used to conduct economic analyses within a program.

- Unique or novel applications of existing economic and financial methodologies, particularly those that are recognized to be outside of mainstream economic practices.

- Broad-scale economic analyses of regulatory programs, such as those required by Congressional mandates (e.g., the Clean Air Act reports to Congress on benefits and costs).

- Stated preference (e.g., contingent valuation) and revealed preference surveys (e.g., recreational travel cost surveys) developed to assist in the economic analysis of a regulation or program.

- National surveys of costs and expenditures for environmental protection (e.g., financial needs surveys, pollution abatement expenditures surveys).

- Economic multiyear research plans developed to assess and advance the state-of-science in economic theory, methodologies or modeling (in particular, the technical feasibility of the plan's components).

- Meta-analyses (i.e., re-analyses of existing published literature and supporting data on the measurement of economic benefits, costs and impacts) developed to assist in the economic analysis of a regulation or program.

Other economic work products also might benefit from peer review, even though they do not exhibit a high degree of complexity or establish an innovative approach. For these, factors such as the potential significance of the analysis for cross-agency or interagency practices or the significance of the issue addressed may make peer review desirable. Examples include:

- Analyses measuring the economic impacts and effectiveness of adopting market-based or economic incentives as regulatory management instruments.

- Technical analyses supporting economic policies established under other government organizations (e.g., economic models used to study transportation, economic development and international trade policies).

External peer reviews can be provided by the SAB's Environmental Economics Advisory Committee, other appropriate outside organizations, or individual, non-EPA reviewers who have expertise in the technical economic issues raised in the economic work product.

3.5.8. Should Economic Analyses Prepared in Support of "Major" or "Economically Significant" Regulations Be Peer Reviewed?

If an Economic Analysis or Regulatory Impact Analysis[21] uses accepted, previously peer-reviewed methods in a straightforward manner, it would not undergo additional peer review. The OMB Peer Review Bulletin specifically exempts Economic Analyses already reviewed through an interagency review process that involves application of the principles and methods defined in OMB Circular A-4.[22] Furthermore, Economic Analyses prepared to support "major" or "economically significant" regulations[23] typically do not utilize innovative or untried economic methods. It is unnecessary to conduct peer reviews of straightforward applications or transfers of accepted, previously peer-reviewed economic methods or analyses (including those published in peer-reviewed journals). Therefore, Economic Analyses that are developed using these procedures do not normally undergo an additional peer review, even those Economic Analyses prepared in support of "major" and "economically significant" rules.

Even when peer review is not required, additional peer input can be beneficial in the development of economic work products for "major" and "economically significant" rules, and this input is encouraged by the OMB Peer Review Bulletin. At present, some peer input of these analyses already is likely to be included as part of the regulatory development process, including input received from other EPA offices represented on the workgroup for the rule, from the Agency's Regulatory Steering Committee, and from the public as part of the public comment process for the rule. There may be, however, added benefit to employing additional peer input procedures, such as actively soliciting input from economists elsewhere in the Agency (through the Economics Forum Steering Committee or the National Center for Environmental Economics), as well as economists from other federal agencies, on the quality and completeness of the Economic Analysis. It is unnecessary to conduct peer reviews of straightforward applications or transfers of accepted, previously peer-reviewed economic methods or analyses, (including those published in peer-reviewed journals).

3.5.9. What Other Social Science Work Products Need Peer Review?

Typically, a social science work product is one that includes empirical, logic-based approaches to answer technical questions about human motivation, human behavior, social interactions and social processes that are relevant to the environmental issues being addressed. The term "behavior" includes overt actions; underlying psychological processes, such as cognition, emotion, temperament and motivation; and bio-behavioral interactions. The term "social" includes socio-cultural, socio-economic and socio-demographic status; bio-social interactions; and the various levels of social context, from small groups to complex cultural systems. Examples of social science work products include analyses

---

[21] The OMB Peer Review Bulletin refers to Economic Analyses as Regulatory Impact Analyses.

[22] OMB. 2003. *Circular A-4, Regulatory Analysis.* http://www.whitehouse.gov/sites/default/files/omb/assets/omb/circulars/a004/a-4.pdf. September 17.

[23] Under Section 3(f)(1) of Executive Order 12866 (58 *Fed. Reg.* 51,735 [Oct. 4, 1993]), "significant regulatory actions" rules are those that may have an annual effect on the economy of $100 million or more or adversely affect in a material way the economy; a sector of the economy; productivity; competition; jobs; the environment; public health or safety; or state, local or tribal governments or communities. The term "major," as defined in the Congressional Review Act (5 U.S.C. § 804(2)), means a rule that has resulted in or is likely to result in: an annual effect on the economy of $100 million or more; a major increase in costs or prices for consumers, individual industries, federal, state or local government agencies, or geographic regions; or significant adverse effects on competition, employment, investment, productivity, innovation or on the ability of U.S.-based enterprises to compete with foreign-based enterprises in domestic and export markets.

and/or evaluations related to such topics as pollution prevention, risk communication, environmental information, environmental justice, quality of life, decision making and public participation.

The following social science work products normally should undergo external peer review:

- Internal Agency guidance for conducting social impact assessments and other community cultural assessments related to different environmental protection approaches, such as community-based watershed protection (heretofore referred to as social assessments).

- New social science methodologies that will serve as a principal method or protocol to conduct social assessments.

- Unique or novel applications of existing social science methods, such as surveys, focus groups, interviews, network analyses, comparative analyses and content analyses.

- New national surveys of values, perceptions and preferences related to environmental protection.

- Innovative research or analyses that address the human dimensions of environmental protection or environmental change in terms of social trends, future predictions and/or behavioral generalizations.

- Social science multiyear research plans developed to assess and advance the state-of-science in social science theory, methodologies or modeling (in particular, the technical feasibility of the plan's components).

3.5.10. Are Regulations Subject to Peer Review?

A regulation itself is not subject to the Peer Review Policy. However, all ISI and HISAs that support a regulatory action should be peer reviewed. The administrative record for the action should include a statement certifying how the peer review provisions have been met (see Appendix D). For discussion of the role of peer review in regulatory development, see Section 1.4.

3.5.11. Should Environmental Regulatory Models Be Peer Reviewed?

In general, the answer is yes. Guidelines for the peer review of environmental regulatory models have been published by the Agency. These can be found on the EPA website under http://nepis.epa.gov/Exe/ZyPDF.cgi?Dockey=P1003E4R.PDF.

4. Peer Review Types and Mechanisms

4.1. Overview

After a planned work product has been categorized as Influential Scientific Information (ISI); a Highly Influential Scientific Assessment (HISA), which is a subset of ISI; or "other," the selection of a peer review approach is needed and involves consideration of many aspects. This chapter outlines the steps for a range of peer review options and discusses the processes and considerations relevant to each (Figure 5). The EPA develops various scientific work products that may be used to support its analyses and decisions. These products vary widely in their complexity and levels of influence. Although much attention is given in this Handbook to influential information, selecting the appropriate type of review mechanism also is important for work products categorized as "other." This chapter, therefore, applies to all products that warrant peer review, not only work products categorized as ISI or a HISA. In addition, although the peer review principles in this Handbook apply to both internal and external peer reviews, the emphasis of this chapter is on options for obtaining external reviews.

4.2. Choosing a Peer Review Mechanism

The preamble to the Office of Management and Budget's (OMB) Peer Review Bulletin[24] notes that

> "… different types of peer review are appropriate for different types of information. Under this Bulletin, agencies are granted broad discretion to weigh the benefits and costs of using a particular peer review mechanism for a specific information product. The selection of an appropriate peer review mechanism for scientific information is left to the agency's discretion."

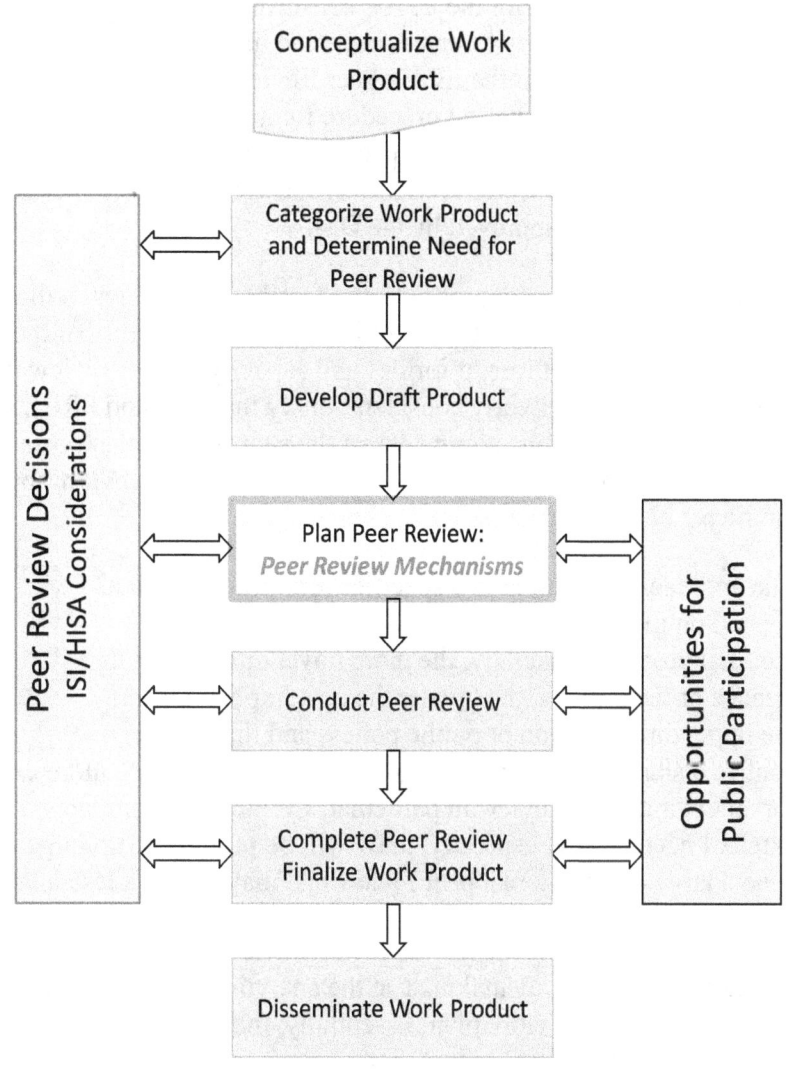

Figure 5. The Peer Review Process: Peer Review Mechanisms

---

[24] OMB. Dec. 16, 2004. Memorandum for Heads of Departments and Agencies, *Final Information Quality Bulletin for Peer Review*. http://www.whitehouse.gov/sites/default/files/omb/memoranda/fy2005/m05-03.pdf.

4.2.1. How Is the Appropriate Peer Review Mechanism Determined?

During the planning of a peer review, the Decision Maker (DM), the Peer Review Coordinator (PRC) and the Peer Review Leader (PRL) may consider several mechanisms for the peer review of a scientific or technical work product. Options range from formal review by EPA colleagues not involved in developing the product (internal peer review or Agency review) to a large and formal panel of subject matter experts from outside EPA (external panel of independent peer reviewers) to a combination of internal and external peer reviews. The peer review effort might be a focused one-time evaluation, or it might encompass several examinations over the course of a product development. Peer review provides the greatest credibility for the EPA's scientific and technical work products when it involves qualified, external independent reviewers; is intensive in its examination; and operates through a formal and transparent process. Per the EPA's Peer Review Policy, external peer review is the approach of choice for all ISI and is the expected procedure for a HISA. Time and resource considerations, however, may impose limitations on the type of peer review performed. If only an internal peer review is planned for scientific and technical work product(s) categorized as ISI or HISAs, the rationale for doing this should be documented and approved by the DM.

Arranging for the most appropriate and feasible peer review will involve a judgment regarding the extent to which the peer review will improve the credibility of the product, as well as consideration of substance, time, resources, priorities and capacity of peer review mechanisms. The PRL should develop a peer review plan for early consideration by the DM (and PRC). For influential work products, including HISAs, public comments on the peer review plan posted on the Science Inventory (SI) (see Section 7.3.4) may lead the Agency to modify the peer review approach, for example, to employ a public panel review process rather than letter reviews.

The approach best suited to a specific work product will depend on the nature of the topic and the intended use of the final product. Generally, the more novel or complex the science or technology, the greater the cost implications of the impending decision or public policy, and the more

> *The mechanism of the peer review should match the importance and complexity of the work product.*

controversial the issue, the stronger the indication is for a more extensive and involved peer review and for an external peer review in particular. Certain work products may lend themselves clearly to extensive external peer review; generally, these will be products with large impacts. Other work products may not need a large-scale external peer review and may utilize a less involved, less resource-intensive review.

It is important to make the choice of peer review mechanism at the time that the work is planned (for products supporting rulemakings, at the analytic blueprint stage) so that peer review costs and time can be budgeted into the work plan. Essentially, the level of peer review should match the impact and complexity of the work product. For example, a Tier 1 or Tier 2 rule under development carries considerable weight and deserves careful handling and attention; therefore, in cases where the Agency has determined that a supporting work product should be peer reviewed, that peer review deserves a commensurate level of care and attention.

Factors that should be considered in selecting a peer review approach include the categorization of the work product (ISI, HISA or other), the availability of internal or external qualified reviewers with the required expertise, whether individual or group advice is desired, and the provision for opportunities for the appropriate level of public participation. Timing and budgetary considerations also may be factors. No single peer review mechanism is likely to work best in all situations; the DM, PRC and PRL should consider, however, the following general guidance:

- For ISI and HISAs intended to support the most important decisions, or for work products that have special importance in their own right, the recommended approach is an internal review followed by an external peer review. Generally, the more complex, novel and/or controversial the product, or the higher impact it is likely to have, the more the DM should consider implementing a peer review involving external experts and providing opportunities for public participation.

- HISAs (a subset of ISI) are expected to undergo rigorous external peer review with opportunities for public participation. When time and resources allow, panels are preferable. External panels usually will be managed by a contractor or conducted by a federal advisory committee (FAC).

- Work products that are less complex, novel or controversial, or that have a lower impact, may be subject to less extensive, less resource-intensive review processes.

- Group discussion among peer reviewers (i.e., panel reviews) can be very helpful in the peer review process because it allows interaction among peer reviewers with different perspectives and expertise. Peer review panels to which the public is invited are more transparent than closed discussions.

- In general, more reviewers are necessary for complex projects (to ensure that expertise from more disciplines is represented) and for controversial topics (to represent differences in scientific perspective within a discipline).

- Strict time constraints, such as a court-ordered deadline, can make a less involved or less formal peer review mechanism imperative. DMs and PRLs should make maximum efforts to ensure that such a process is systematic and objective.

- Reviews of products from remediation and other programs may be tied to litigation; the Office of General Counsel (OGC) or the Office of Regional Counsel (ORC) should be consulted regarding any restrictions to be aware of before deciding what peer review mechanism to use.

4.2.2. What Are Some Examples of Internal Peer Review Mechanisms?

The following are examples of internal peer review mechanisms:

- Individual letter review by independent EPA experts (e.g., a review by Office of Research and Development [ORD] experts of a draft article on benchmark dose completed by a program office).

- *Ad hoc* panel of independent EPA experts (e.g., an independent internal workgroup convened to review the science supporting the possible classification of a chemical as a carcinogen).

- Technical review by scientists in an EPA laboratory, typically conducted by letter (e.g., an initial review of the risk assessment for a regional incinerator by agency scientists), prior to submission to a journal.

4.2.3. What Are Some Examples of External Peer Review Mechanisms?

Examples of external peer review mechanisms include the following:

- Review of a journal manuscript by a refereed scientific journal.

- Letter review by individual independent experts from outside the Agency.

- *Ad hoc* panel of independent non-EPA experts convened for review and discussion, with each panelist submitting his/her comments separately.

- Review by an established FAC (e.g., a review of an Integrated Scientific Assessment document for a criteria air pollutant by the Clean Air Scientific Advisory Committee [CASAC]).

- Agency-appointed special board or commission (e.g., a review of the risk assessment methodology prepared by the Clean Air Act Commission on Risk Assessment). OGC should be consulted to determine whether the Agency has specific statutory authority to establish and finance the activities of a board or commission that would perform governmental functions and whether the Federal Advisory Committee Act (FACA) would apply to the board or commission.

- Review by the National Academy of Sciences (NAS) under a contract with EPA.

There are other bodies that may provide external commentary on Agency work products but are not considered peer review mechanisms, such as the following:

- Interagency committees (e.g., a review of prospective research plans by the Committee on the Environment, Natural Resources, and Sustainability, coordinated by the White House).

- Committees convened by another federal agency or government organization (e.g., a review of the Dioxin Reassessment by the Health and Human Services Committee to Coordinate Environmentally Related Programs).

- Reviews initiated by nongovernmental groups (e.g., a Society for Risk Analysis review of cancer guidelines).

4.3. Mechanism: Journal Peer Review

Peer review of journal articles performed by a credible, refereed scientific journal contributes to the scientific and technical credibility of the reviewed product. Generally, EPA considers peer review by such journals as adequate for reviewing the scientific credibility and validity of the findings (or data) in that article and, therefore, a satisfactory form of peer review.

Prior to submitting an article to a journal for peer review, EPA employees are encouraged to have the article internally peer reviewed. Articles also may need examination in accordance with any organizational clearance procedures, especially when the author includes EPA as their affiliation. For EPA employees, Conflict of Interest (COI) law and policy also will apply.

The OMB Peer Review Bulletin does not apply to journal articles because such publications do not contain findings or conclusions that represent the official position of the Agency (i.e., they are categorized by the Agency as "other"). Therefore journal articles must have the appropriate disclaimer that the work represents the views of the author(s) and not those of the Agency (e.g., "The views expressed in this paper are those of the authors and do not necessarily reflect the views or policies of the U.S. Environmental Protection Agency").

4.4. Mechanism: Letter Reviews

Generally, letter reviews by several experts will be more expeditious and less expensive than convening a panel. Letter reviews by individual experts are more appropriate when a work product is not controversial, covers only a few disciplines, or when premature disclosure of a sensitive report to a public panel could cause harm to government or private interests. The letter review process may include a public comment period on the draft Agency document, with comments received from the public being shared with the peer reviewers. There also are multistage processes in which letter reviews may be conducted prior to the release of a work product for public notice and comment, followed by a formal panel review. These multistage processes are particularly valuable for highly complex and multidisciplinary products, especially those that are novel or precedent-setting.

Letter peer reviewers are selected primarily according to their areas of expertise, knowledge, skills and experience. They are evaluated for independence, potential COI and appearance of a loss of impartiality (see Chapter 5) before being selected for a letter review. If letter peer reviewers will be compensated using a purchase order or contract mechanism, the PRL should work with the Contracting Officer (CO) to develop an appropriate task statement or scope of work. Guidance is provided in Section 4.6. If letter peer reviewers are not to be compensated, they will need to sign a Gratuitous Services Agreement for Peer Review, as discussed in Section 4.6.7.

4.5. Mechanism: Panel Reviews

When time and resources permit, panels are preferable for influential products because they tend to be more deliberative than individual letter reviews and the reviewers can help inform one another. Panels are valuable when the work product is complex and multidisciplinary. Panel peer review meetings may be open to the public, with opportunities for public comment. Peer review panels that include EPA experts do not constitute external peer review.

The Agency may organize internal peer review panels composed of independent EPA experts or a mix of EPA experts and experts from other federal agencies. If Agency-organized panels include nonfederal experts, the provisions of the FACA may apply (see Section 4.7.5).

External peer review panels, in most cases, will be managed under a peer review contract (see Section 4.6) or conducted by a chartered FAC (see Section 4.7). Another option for obtaining external panel peer review is for the Agency to contract with the NAS (see Section 4.8).

4.6. Peer Review by Contractors

4.6.1. Can the Agency Use a Contractor to Obtain Peer Review Services?

Yes, the Agency can use a contractor to obtain peer review services.[25] Peer review services are "advisory and assistance services," as defined in Federal Acquisition Regulation (FAR) 2.101. It should be noted that these types of services require special approvals and management oversight. Approval

---

[25] If EPA manages or controls a group convened by a contractor, the FACA may apply.

levels for advisory and assistance services are located in the in Subsection 1.6.1 of the EPA Acquisition Guide (EPAAG) available at http://oamintra.epa.gov/node/521.

Typically, peer review services would be available under a "mission contract," that is, a contract with a broad scope covering a variety of services. It also is possible to have a contract or purchase order solely for peer reviews (see Section 3.4). A contractor assisting the Agency in the development of a work product, however, should not be used to provide peer review services for that same work product.

The Agency may obtain peer review services through a contract or purchase order. Contracts or purchase orders may be used to obtain both letter and panel review services, and this guidance applies to both. A contract is awarded if the cost is more than the simplified acquisition threshold ($150,000 in fiscal year 2015). If the cost is $150,000 or lower, then a purchase order typically is issued. For assistance in preparing the necessary pre-award contract documents, Chapter 7 of the EPAAG and the appropriate contracting office should be consulted.

For assistance in preparing simplified acquisition packages for purchase orders, the Office of Acquisition Management has a guide called *SAME: Simplified Acquisition Made Easy*, which is available on the intranet at http://oamintra.epa.gov/files/OAM/sapsEasy.pdf.

4.6.2. How Does the Peer Review Leader Write a Statement of Work (SOW) for Peer Review Contracts?

The SOW should specify clearly that the contractor is responsible for preparing peer review evaluations and should set forth guidelines for the peer review of scientific or technical documents. The contractor may perform the peer review with appropriate contractor staff, subcontractors or consultants.

> *Contracts may be used to obtain both letter and panel review services.*

Any guidelines needed to ensure the soundness and defensibility of peer reviews should be developed by the EPA office and made part of the contract. The contractor then would ensure that the peer reviews adhere to the guidelines.

If the charge questions are known prior to the issuance of a solicitation for a contract, or prior to the issuance of a tasking document under an awarded contract, the CO can incorporate the charge questions directly into the SOW for the contract or tasking document. Otherwise, the charge questions would be provided to the contractor in a separate tasking document or technical directive.

The SOW must specify the full range of desired services. Unless the prime contractor is clearly tasked with responsibility for performing peer reviews and delivering peer review comments or a peer review report, individual peer reviewers' fees and associated travel expenses are not payable under the contract. If the SOW calls for the preparation and delivery of comments or an evaluation, as well as specifying a meeting with the Agency and other peer reviewers as part of the peer review, payment is appropriate. The peer reviewer's attendance at the meeting then would be part of contract performance. The prime contractor, rather than EPA, must select the peer reviewers, although the terms of the contract may specify qualifications for peer reviewers and EPA may review the qualifications of peer reviewers the contractor proposes to hire to ensure they meet the established qualifications. Example SOWs are presented in Appendix E.

4.6.3. Can the Agency Select Peer Reviewers When Using a Contractor-Managed Peer Review?

When using a contractor-managed peer review, the Agency cannot select peer reviewers.[26] When a contractor is managing a peer review (either by panel or letter) for the Agency, the prime contractor is responsible for selecting who will perform the peer review. Interfering in this process may be a violation of federal and Agency acquisition regulations. Specifically, it may constitute directed subcontracting.

The EPA can establish qualifications for peer reviewers. The Agency should not be involved, however, in the selection of individual peer reviewers and should avoid commenting on the contractor's selection of peer reviewers other than to determine whether the reviewers, once selected, meet the qualifications established, including compliance with contract requirements pertaining to COI. The EPA may identify, however, a pool of qualified peer reviewers for the prime contractor to consider. The candidates should be listed in alphabetical order and, to avoid directed subcontracting issues, the list generally should include more individuals than the number required for the review.

If a list is provided, it should be noted on the list that it is a suggested list and other qualified candidates may exist who are not on the list. This is to prevent the impression that the prime contractor can choose only someone on the list. The prime contractor is required to include several COI clauses substantially similar to the COI clauses included in the primary contract in its subcontracts with the peer reviewers.

4.6.4. How Is the Panel Formed When a Contractor Manages a Panel Peer Review for ISI or HISAs?

In March 2013, the Science and Technology Policy Council (STPC) approved a process to enhance the transparency and the EPA's oversight of panel peer reviews of ISI and HISAs when the reviews are managed by contractors (see the EPA's *Conflicts of Interest Review Process for Contractor-Managed Peer Reviews of EPA HISA and ISI Documents*, http://www2.epa.gov/osa/conflicts-interest-review-process-contractor-managed-peer-reviews-epa-highly-influential). Under this process, EPA will publish a "Call for Experts" in the *Federal Register* to identify the types of expertise needed, announce the availability of the document to be reviewed or provide a brief synopsis of the document, direct the public and stakeholders to submit nominations of potential peer reviewers to the contractor, and allow a minimum of 3 weeks for the public to nominate expert candidates. At the same time, the contractor will use traditional techniques to identify additional qualified candidates in the disciplines identified by EPA. The contractor will screen all nominees (including those submitted by the Agency and the public) for expertise and potential COI. Based on the information collected by the contractor, the contractor will develop a list of potential peer reviewers. This list of potential peer reviewers will be published for public review and comment.

The process for contractor managed panels also provides for more direct interaction between EPA and the contractor in addressing actual or potential COIs. All prospective reviewers for contractor-managed panel reviews are evaluated for independence, COI and an appearance of a loss of impartiality and are required to complete COI disclosure forms. Among other things, these forms require prospective reviewers to disclose to the contractor certain financial interests and answer questions regarding connections to the work product being reviewed. An example COI Statement form is included in Appendix J. In addition, the CO and the Contracting Officer's Representative (COR), in consultation

---

[26] If EPA were to select the reviewers for a contractor peer review involving group advice of the peer review panel, FACA may apply because EPA would be exerting control over the panel.

with the EPA Science Advisor (or his or her designee), will discuss with the contractor the process used by the contractor to identify and address COI, ensure that the contractor and prospective reviewers are in compliance with COI requirements in the contract, and provide input on any issues concerning potential conflicts.

4.6.5. What Are Some Management Controls for Peer Review Contracts?

Contract management controls are designed to ensure the following:

(1) The contractor does not perform inherently governmental activities (IGA).

(2) The contractor and the contractor's work is free from COIs or conflicts can be appropriately avoided, neutralized or mitigated.

(3) If provided to the contractor, confidential business information (CBI) or other confidential/sensitive information is appropriately safeguarded.

(4) Improper relationships with contractor employees and subcontractors are avoided.

Each of these concepts is discussed in the sections that follow.

4.6.5.1. What Are Inherently Governmental Activities and What Management Controls Prevent Contractors from Performing Them?

Agency regulations and FAR prohibit contractors from performing IGA. OMB Policy Letter 11-01 (76 *Fed. Reg.* 56,227, Sept. 12, 2011) defines "inherently governmental activities" as activities that are so intimately related to the public interest as to mandate performance by government personnel. These activities require the exercise of substantial official discretion in the application of government authority and/or in making decisions for the government.

With contracts for peer review services, the Agency is seeking only a contractor's recommendations, advice or analysis of a document, not a determination of whether the document is acceptable for the EPA's purposes or what the policy that the document supports should be. Determining Agency policy is an IGA. EPA officials make the official Agency decision regarding acceptability and/or quality of the document. To ensure that Agency officials are not influenced improperly by the recommendations in the peer review, the contract should include management controls. One possible control would be to direct the peer reviewers to submit with their evaluations or comments a description of the procedures used to arrive at their recommendations, a summary of their findings, a list of sources relied upon and clear and substantiated identification of the methods and considerations upon which their recommendations are based. To the extent possible, the contract should set forth any guidelines or criteria for performance of the peer review. Agency officials should document their evaluations of the quality and validity of the peer review, including a clear record of their review of the contractor's work and documentation that Agency personnel made the final decisions. Such records of review could include notes from reviews of draft and final documents by EPA personnel and minutes from progress meetings with contractors.

4.6.5.2. What Are Management Controls for Conflict of Interest?

To identify and avoid, neutralize or mitigate actual or potential COI, the contract should include controls. Inclusion of Agency-developed personal and organizational COI clauses in the contract or purchase order is critical when procuring peer review services. Usually, the CO will include COI solicitation provisions and contract clauses as a matter of course without involvement by the EPA Project Officer. As a safeguard, the COR should:

- Section 9.5 of the EPAAG, which provides guidance and procedures for addressing and documenting organizational COI. Project Officers also should review the Office of Acquisition Management's News Flash Notice titled "Evaluating Conflict of Interest Issues Pre-Award" (August 11, 2006) (available at http://oamintra.epa.gov/node/47?q=node/80).

- Highlight the COI requirements in the SOW for the procurement of the peer review services. In particular, the COR should ensure that the peer review "COI Evaluation for Task Orders/Work Assignments" clause is included in the contract (see Appendix J for the text of the clause). Responses to the questions included in the clause are considered confidential in accordance with applicable laws and regulations, and they are used to identify any potential COI.

- Review the solicitation/contract to ensure that other appropriate COI clauses have been included, particularly EPA Acquisition Regulation clauses 1552.209-70, Organizational Conflict of Interest Notification; 1552.209-71, Organizational Conflict of Interest; 1552.209-72, Organizational Conflict of Interest Certification; and 1552.209-73, Notification of Conflicts of Interest Regarding Personnel.

- Work with the CO to develop contract-specific language regarding the peer review to assist the contractor in identifying actual or potential COI that might impair the objectivity of peer reviewers. For example, the peer review COI Evaluation clause advises contractors to consider the questions and issues listed in Exhibit 4 when determining if a proposed peer reviewer may have an actual or potential COI or bias.

Peer reviewers appointed through a contract mechanism, either by contracting directly with EPA or by being selected by a peer review contractor, are not government employees. Accordingly, the COI statutes and ethics regulations that apply to Regular Government Employees (RGEs) and Special Government Employees (SGEs) do not apply to them. "Appearance" issues with respect to experts hired through a contract mechanism, however, are addressed under the FAR definition of "organizational conflict of interest" (FAR 2.101). Among other things, the definition includes situations in which "because of other activities or relationships with other persons, a person is unable or potentially unable to render impartial assistance or advice to the Government" (FAR 2.101).

In addition, FAR 3.101 advises that COs should strictly avoid even the appearance of a COI in government-contractor relationships. When evaluating "appearance" issues with respect to experts hired under a contract mechanism, the CO may consider facts and circumstances similar to those that a PRL might consider when evaluating "appearance" issues for SGEs and RGEs. These include: the nature of the relationships involved, financial considerations, prior statements, testimony, work related to the subject matter of the peer review and other factors bearing on an expert's impartiality (see Section 5.3.7).

Exhibit 4. Questions and Issues Contractors Should Consider When Determining if a Proposed Peer Reviewer May Have an Actual or Potential COI or Bias

- o The sources and nature of any compensated and uncompensated employment of the panel member and their spouse (obtained from a brief description of the work), including any government service, for the preceding 2 years.

- o The sources of research support and project funding, including from any government source, for which the panel member served as the Principal Investigator (PI), Significant Collaborator, Project Manager (PM) or Director during the preceding 2 years. For the panel member's spouse, a general description of research and project activities in the preceding 2 years.

- o The compensated consulting activities of the panel member during the preceding 2 years, including the names of clients if the compensation provided 15 percent or more of the member's annual compensation. For the panel member's spouse, a general description of consulting activities for the preceding 2 years.

- o The sources of compensated expert witness activities of the panel member and a brief description of the issue and testimony during the preceding 2 years. For the panel member's spouse, a general description of expert testimony provided in the preceding 2 years.

- o The assets—including stocks, bonds, real estate, business, patents, trademarks and royalties—of the panel member, their spouse and dependent children. Specifically, the financial holdings that collectively had a fair market value greater than $15,000 at any time during the preceding 2-year period (excluding, for example, well-diversified mutual funds, money market funds, treasury bonds and personal residences).

- o The liabilities more than $10,000 owed by the panel member, their spouse and dependent children at any time in the preceding 1 year (excluding, for example, a mortgage on a personal residence, home equity loans and automobile and consumer loans).

- o A brief description of any public statements and/or positions of the panel member on, or closely related to, the matter under review.

- o A brief description of any previous involvement of the panel member with the development of the document (or review materials) that the individual has been asked to review (including previous peer reviews).

- o A brief description of any other information that might reasonably raise a question about an actual or potential personal COI or bias, including any financial benefit that might be gained by the panel member (or anyone whose interests are imputed to the panel member) as a result of the outcome of the review.

The CO, not the contractor, has the authority under the FAR and EPA Acquisition Regulations to determine whether "appearance" or other COI issues exist. When evaluating "appearance" and other COI issues, however, the CO may seek the advice or expertise of others, such as the Project Officer, CORs, Agency technical and subject matter experts, the EPA Science Advisor or his/her designee and

OGC. The CO also has the authority to determine whether "appearance" and other COI issues can be appropriately avoided, neutralized or mitigated.

4.6.5.3. What Management Controls Protect Confidential Business Information/Privacy Act-Protected Information and Other Privileged/Sensitive Information?

When peer reviewers are not employees or contractors/subcontractors of the U.S. Government, it is unlikely that EPA will have authority to give reviewers access to CBI or other protected or sensitive information in the absence of consent for such disclosure by the CBI submitter or other interested parties. Therefore, all documents provided to nonfederal reviewers must be screened for information claimed as CBI or other protected information.

Even where business information has not been explicitly claimed as CBI, if it is of a kind that the submitter might be expected to object to its release, prior to release the submitter must be asked whether it wants to assert a claim, unless the submitter previously has been informed that failure to assert a CBI claim may result in disclosure without notice, as consistent with 40 C.F.R. § 2.203. If the contractor should have access to CBI for the peer review, the CO must be notified so that the appropriate clauses can be included in the contract or purchase order. These clauses will identify clearly any required procedures or processes prior to release of any protected information, including any requirements for confidentiality agreements, as well as limits on use and disclosure of the data by contractor personnel.

In general, materials provided by EPA to the contractor, or generated by the contractor or subcontractors during performance of the contract, should be protected from release until EPA determines the information is not entitled to confidential treatment. Appropriate contract clauses (e.g., EPA Acquisition Regulation §1552.227-76, "Project Employee Confidentiality Agreement"; FAR 52.227-17, "Rights in Data—Special Works") should be included in the contract and subcontracts with individual reviewers to ensure that such materials are not copied, shared or otherwise distributed or forwarded to others, except as provided for in the contract or as authorized in writing by the CO. The contractor is free to consult with colleagues (unless otherwise directed) on technical issues raised in the draft report but not to share the draft report itself (see Section 6.2.5).

4.6.5.4. What Management Controls Prevent Improper Personal Services?

Contractor employees must not be treated as EPA employees unless statutory authority exists to engage the contractor employee in personal services contracts. For additional information, program officials should consult EPA Order 1901.1A, *Use of Contractor Services to Avoid Improper Contracting Relationships* (http://intranet.epa.gov/ohr/rmpolicy/ads/orders/1900-1achg2.pdf).

To avoid these improper relationships, the SOW should be well-defined and should set forth a detailed description of the work to be performed independently, including the manner in which it will be evaluated. The SOW should state what work is to be performed, not how the work is to be performed. Technical direction may be used to clarify ambiguous provisions to ensure efficient and effective contractor performance and is not considered supervision or assignment of tasks.

4.6.6. How Is Peer Reviewer Travel Handled With Contracts or Purchase Orders?

Funds obligated on a contract or purchase order are available to pay for the costs of producing the peer review, including the travel costs and fees of the peer reviewer, provided that the SOW contains language that ensures that the agreement is for providing a service or product rather than simply paying for peer reviews' travel.

The EPA may acquire peer review services through purchase orders issued directly to peer reviewers or through contracts with companies that manage and provide the peer review services. By issuing a purchase order or awarding a contract for peer review services, EPA may pay not only for the peer review services/comments, but also for travel necessary for the peer reviewer's participation in a meeting with the Agency and other reviewers to discuss comments. The scope of work of the contract, however, must require the contractor or individual peer reviewer, as appropriate, to perform the peer reviews and produce peer review comments or a peer review report, and to discuss a specific peer review work product with the Agency and/or with other peer reviewers in person. Participation in a meeting to discuss a peer review work product then would be part of the contractor's performance. While EPA may use GSA's per diem and meals and incidental allowances as a basis for negotiating travel costs, the terms of the contract or purchase order should not imply that peer reviewers receive travel reimbursement under the federal travel regulations. Under these circumstances, the contract may serve as the mechanism to pay for peer review services and associated travel expenses to provide comments to EPA.

4.6.7. What Are Gratuitous Services Agreements for Peer Review (GSAPR)?

A Gratuitous Services Agreement for Peer Review (GSAPR) is a written agreement between an authorized EPA official (PRL) and a nonfederal peer reviewer under which the peer reviewer agrees to provide EPA with a report, analysis or similar work product without charge to the Agency. GSAPRs are used when EPA has not appointed a peer reviewer as an unpaid expert or consultant under 5 U.S.C. § 3109 and EPA Order 3110.4A4 "Employment of Experts and Consultants."

The Antideficiency Act (31 U.S.C. § 1342) prohibits the Agency from accepting uncompensated "voluntary" services unless specifically authorized by law.

Generally, improper voluntary services are those provided "for free" to the EPA either for work that must be performed by a federal employee or another individual entitled to statutory compensation or without a written agreement in advance that protects the EPA from future claims for compensation for services rendered. In contrast, under appropriate circumstances, the Agency may accept "gratuitous" services. Gratuitous services are services rendered without compensation under a formal written agreement in which the service provider explicitly agrees that the services will be provided free of charge to the government and that no future claim related to the services will be made. Such agreements must be signed by the service provider before the services are performed. For situations concerning state employees, see Section 5.2.9.

A proper GSAPR must be signed and include a compensation/claim waiver and appropriate terms and conditions that address deliverables, schedules, COI, CBI and other issues relevant to the peer review services provided. It must also include a statement that the peer reviewer understands that he or she will not be considered an employee of the Government for any purpose. The PRL should consult OGC for appropriate compensation/claim waiver language and to ensure that appropriate provisions are included in the agreement to protect the agency's interests.

GSAPRs also are subject to competition requirements, although if EPA's estimate of the value of the services is less than the prevailing micro purchase limit (e.g., $3,000 for Fiscal Year 2014), the competition requirements are relaxed substantially. The PRL should consult a CO when the use of GSAPRs is being considered.

4.7. Peer Review by Federal Advisory Committees

4.7.1. What Is the Role of Federal Advisory Committees in Peer Review?

EPA has a number of scientific and technical advisory committees composed of non-EPA experts who provide advice and peer review to the Agency. The FACA (5 U.S.C. § App. 2) requires that these groups of advisors be fairly balanced in terms of points of view represented for the function to be performed by the committee. Meetings are announced in advance and are open to the public except under limited circumstances (i.e., if the meeting falls within exceptions under the Government in the Sunshine Act, 5 U.S.C. § 552b). All materials presented to and prepared for or by the committees are available to the public, usually on committee Web pages on EPA website. In addition, the FACA requires that the public have an opportunity to provide written comments, and in most cases, advisory committees schedule time at meetings to hear oral public comments on the technical work at hand.

The EPA has more than 20 formally established FACs, but not all are set up to conduct scientific peer review (e.g., some committees are established to provide policy advice to the Agency, rather than scientific and technical review). The scope of work of each advisory committee is set out in its charter, a formal document filed with Congress when the committee is established and renewed every 2 years. Scientific and technical advisory committees are composed of members who are appointed because of their expertise, rather than as a representative of an organization or interest group. Committee members on scientific and technical FACs serve as SGEs or non-EPA RGEs and are subject to ethics laws and regulations that apply to employees of the Executive Branch (see Section 5.3). If no existing FAC has the appropriate expertise, a new FAC could be established to conduct the peer review.

Because of FACA requirements for open meetings, transparent deliberations, formal opportunities for public participation and publicly available records, scientific FACs provide an external peer review mechanism that meets the provisions in the OMB Peer Review Bulletin for peer review of HISAs.

The Science Advisory Board (SAB) Staff Office, in the Office of the Administrator, provides administrative and technical support to two scientific advisory committees: the EPA SAB and the CASAC. When either of these committees is the mechanism for obtaining external peer review, the SAB Staff Office budgets for, plans and manages the peer review meetings. The SAB Staff Office selects peer reviewers after a public nomination and comment process and after screening for ethics issues such as potential COI or an appearance of a loss of impartiality. The SAB Staff Office also announces committee meetings in the *Federal Register* and on EPA committee websites, prepares detailed meeting minutes, transmits the EPA's charge and review materials to the committee and provides support to the committee in preparation of the advisory report to the EPA Administrator. To maintain the independence of the peer review process, the SAB Staff Office does not draft the EPA charge or prepare the Agency response to the peer review. The SAB Staff Office also does not enter data into the SI.

> *When external peer review is conducted by the SAB or the CASAC, the SAB Staff Office performs many—but not all—of the functions of the Peer Review Leader.*

4.7.2. When Is It Appropriate to Seek Peer Review from EPA's Science Advisory Board?

The EPA's SAB is a statutorily established committee with a broad mandate to provide advice and recommendations to the Agency on scientific and technical matters. The SAB considers requests for advice and peer review from across the Agency as part of an annual process, initiated by a request from the Deputy Administrator (DA) to the EPA's senior leadership to identify requests for review by EPA

FACs in the coming year. In a complementary semiannual process coordinated by the EPA Office of Policy, the SAB also considers review of science supporting major planned Agency actions (Tier 1 and Tier 2 actions) that are in the pre-proposal stage.

HISAs or other scientific work products associated with highly visible or controversial environmental issues, or products that include novel scientific methods or approaches, are most suited to review by the SAB.

Much of the SAB's peer review work is done using *ad hoc* panels formed to review specific EPA draft technical products. All SAB panels provide advice through the chartered SAB, which is composed of approximately 50 nationally renowned scientists, engineers and economists. The SAB reports directly to the EPA Administrator. For more information on the SAB, see http://www.epa.gov/sab. Information on the process to request peer review and advice from the SAB is provided in Appendix F.

4.7.3. What Other Federal Advisory Committees Can Provide Peer Review?

In addition to the SAB, EPA has other scientific advisory committees that provide advice and peer review for specific EPA offices. For example, the Board of Scientific Counselors advises ORD on the operation and management of its research programs; the CASAC provides advice on the scientific and technical aspects of air quality criteria and standards; and the Federal Insecticide, Fungicide and Rodenticide Act Scientific Advisory Panel (SAP) provides advice on science issues associated with the EPA's pesticide-related regulatory actions. For a full list of EPA scientific and technical advisory committees, see Appendix G.

4.7.4. How Is Travel Handled for Advisory Committee Members?

Members of the SAB, SAP and other scientific or technical FACs usually are appointed as SGEs. The term "Special Government Employee" is defined in 18 U.S.C. § 202(a) as an officer or employee of an agency who performs temporary duties, with or without compensation, for not more than 130 days in a period of 365 days, either on a full-time or intermittent basis.

Travel and per diem expenses of experts hired as SGEs for peer review may be paid only through the issuance of invitational travel orders (5 U.S.C. § 5703). These invitational travel and per diem expenses should be charged to an appropriate EPA travel account. The Federal Travel Regulations govern the invited traveler's reimbursement. It is not appropriate to reimburse travel or per diem expenses of advisory committee members (SGEs) through a contract.

4.7.5. When Does the Federal Advisory Committee Act Apply to Other Peer Review Mechanisms?

In addition to formally established (chartered) FACs, other groups of peer reviewers may become subject to FACA requirements if they meet all the following criteria:

- Are established, controlled or managed by EPA;

- Include one or more individuals who are not full-time or permanent part-time federal employees; and

- Are intended to, or do, provide group or collective, rather than individual, advice.

EPA-run peer reviews that were not intended originally to be subject to FACA requirements may become subject to them if they exhibit all of the above characteristics. Similarly, if EPA personnel begin to manage or control a contractor-managed peer review, the process may become subject to FACA (see Section 4.7.7). Questions concerning the applicability of the FACA to peer review meetings should be addressed to FACA experts in the Cross-Cutting Issues Law Office of OGC or the appropriate ORC.

4.7.6. When Are EPA-Run Peer Reviews Not Subject to FACA Requirements?

If EPA conducts a peer review by obtaining advice from individual peer reviewers and not for the purpose of obtaining a peer review product from the group as a collective or consensus body, the peer review, in most cases, would not be subject to FACA requirements. When peer review participants provide only their own views or

> *As a general matter, letter reviews that seek individual views or comments are not subject to the requirements of the FACA.*

recommendations and do not vote, develop consensus recommendations to EPA, or use any other means of developing group advice, the FACA does not apply. When referring to the recommendations of the individual reviewers, EPA should **not** characterize these recommendations using such phrases as "the peer reviewers all agreed" or such terms as "collective" or "consensus." As a general matter, letter reviews that seek individual views or comments are not subject to FACA requirements.

In addition to ensuring that peer reviewers only provide comments as individuals, EPA officials can lessen the potential for a challenge under the FACA by conducting the peer review in an open transparent manner (e.g., by seeking a balance of points of view among the peer review participants, allowing interested members of the public to attend peer review meetings, allowing public comment, and ensuring that the public has access to all the peer review materials).

Non-FACA peer review meetings may be advertised publicly through the *Federal Register* and/or other avenues (e.g., the Web, local newspapers and mailing lists). These notifications should provide the public with useful information and a point of contact concerning the peer review. Notice of such meetings, however, should make clear that the meeting is not subject to FACA requirements.

4.7.7. How Does the Agency Ensure That Contractor-Managed Peer Reviews Do Not Inadvertently Invoke FACA Requirements?

Under the current case law, committees (or other peer review groups) established, controlled and managed by an outside organization (such as by an EPA contractor) to provide that outside organization with advice and recommendations (that will be submitted eventually to EPA as a contractor report) are not subject to FACA requirements. Although the FACA should not apply to contractor-managed peer reviews, EPA personnel can do things that might invoke the FACA inadvertently.

The following are considerations that EPA personnel should be aware of when a contractor manages a peer review (e.g., letter review or panel) for EPA:

- The outside party's peer review may be subject to FACA requirements if EPA establishes, manages or controls the peer review group (e.g., EPA selects or rejects peer reviewers, sets the agenda, runs the meeting, or provides funds directly to the peer reviewers). The EPA can make suggestions to the contractor but to avoid triggering the FACA, the contractor must be free to accept or reject these suggestions.

- EPA should not provide contractors with a draft agenda or suggested format for meetings. EPA contractors should manage and control the process, including running any meetings.

- At the request of the EPA contractor, EPA may provide a briefing to the peer reviewers (e.g., in a conference call with the contractor on the line) on the history or background of the development of the document. EPA should provide only technical or background information and not use the call to manage the contractor's peer review group. Not only should the contractor be on the line, but it should be very clear to all participants that the contractor is in charge of the call. The contractor, not EPA, should invite individuals to participate, make all administrative arrangements, conduct the meeting and control the agenda.

- EPA employees may attend the peer review panel meetings, but they may not control the meeting. The contractor may call on them to speak when appropriate, but EPA personnel should limit their participation to answering questions to provide technical and/or background information.

- Because the FACA does not apply when a contractor establishes, controls and manages a peer review, the contractor does not need to avoid terms such as "collective" or "consensus" when reporting agreement among its peer reviewers.

- EPA may provide comments to the contractor on the contractor's peer review report only to the extent that the Agency is verifying that the contractor has satisfactorily completed the report in accordance with the work assignment. EPA should not attempt to make changes in the contractor's conclusions; this would compromise the independence of the peer review conducted by the contractor.

4.8. Peer Review by the National Academy of Sciences

The NAS is a private, nonprofit society of distinguished scientists established by Congress to provide independent, objective advice to the Nation on science and technology matters. NAS review of an Agency work product may be most suitable for significantly controversial or high-visibility products or when required by legislation.

When EPA wishes to obtain peer review services from the NAS, usually through the National Research Council (NRC), the Agency works with NAS staff to develop a set of charge questions (the "statement of task") and to define the duration and cost of the study. Once the statement of task and budget are approved by the NRC Governing Board, the Agency has no control over the conduct of the peer review. Members of the peer review committee are selected by the NAS to provide the appropriate range of expertise and a balance of perspectives. All members are screened for COI in keeping with the NAS Policy on Committee Composition and Balance and COIs.[27]

Like FACs, NAS/NRC committees seek public nominations and comment on peer reviewers and seek to ensure that committees are fairly balanced for the functions to be performed. Unlike FACs, however, NAS/NRC committees conduct fact-finding in public, but deliberate in private.

Official reports from the NAS are generally presumed not to require additional peer review.

---

[27] National Academy of Sciences. 2003. *Policy on Committee Composition and Balance and Conflicts of Interest for Committees Used in the Development of Reports*. Washington, D.C.: The National Academies Press. http://www.nationalacademies.org/coi/bi-coi_form-0.pdf.

5. Peer Reviewer Qualifications and Selection

5.1. Overview

As part of the peer review process, the Agency (or the contractor managing the peer review) must select peer reviewers who have technical expertise in the subject matter that is needed to answer specific charge questions (Figure 6). For this reason, it is important to have a draft or final charge before selecting peer reviewers. These reviewers must not only be subject matter experts, but also must be independent and free from ethics issues such as potential conflicts of interest (COIs) or an appearance of a loss of impartiality (see Sections 5.3.4 and 5.3.7) so that the integrity of the peer review is not brought into question. The rules for evaluating ethics issues of peer reviewers vary depending on the peer review mechanism, but in all cases, adherence to ethical standards is important to ensure that the Agency receives objective, informed and relevant advice through peer review of its work products. Depending on the peer review mechanism chosen, the peer reviewers may be contractors, subcontractors or permanent or intermittent federal employees.

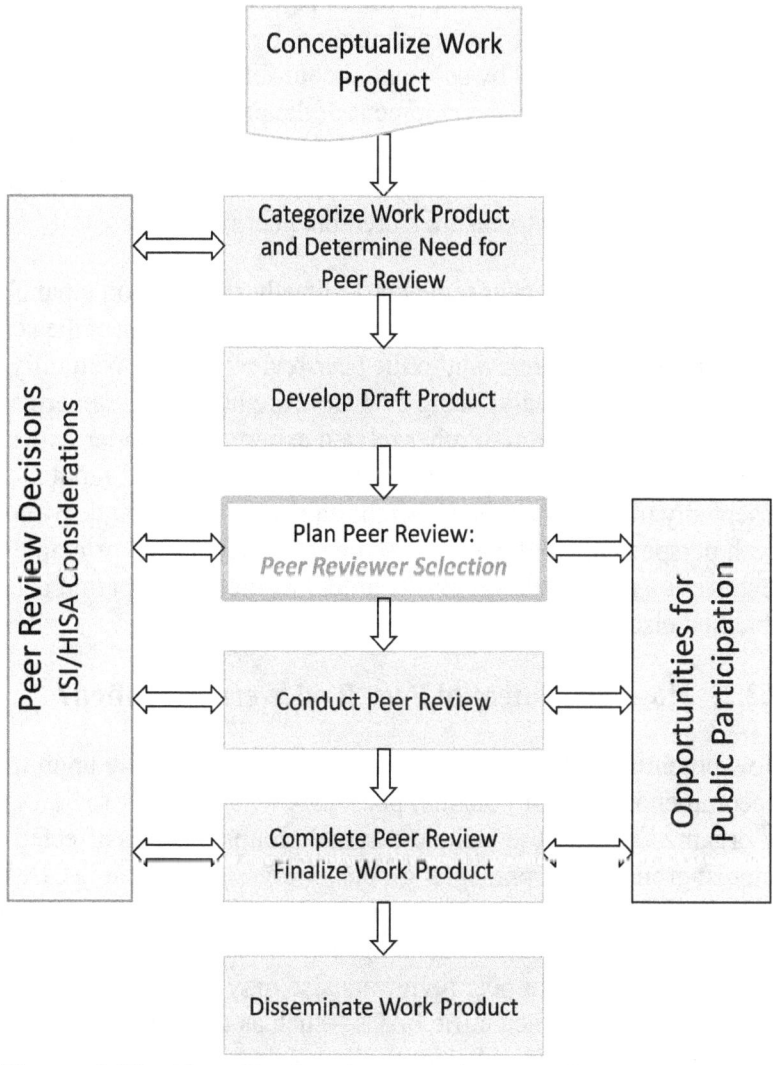

Figure 6. The Peer Review Process: Peer Reviewer Selection

Internal peer reviews can be conducted by independent experts from within EPA, either individually or as *ad hoc* peer review panels. External peer reviews can be conducted by individual experts or panels of experts who are Regular Government Employees (RGEs) at Executive Branch departments or agencies other than EPA, experts appointed to EPA as Special Government Employees (SGEs) pursuant to 18 U.S.C. § 202(a), or experts hired through a contract mechanism. External peer review panels can be convened through a contract mechanism under which EPA uses a contractor who selects the peer reviewers or by a federal advisory committee (FAC) organized pursuant to the Federal Advisory Committee Act (FACA). Lastly, peer reviews may be conducted by outside organizations such as the National Academy of Sciences (NAS).

5.2. Finding Peer Reviewers

5.2.1. What Are the Important Qualifications for Peer Reviewers?

The first consideration in selecting peer reviewers is expertise (i.e., whether the candidates have the knowledge, skills and experience necessary to perform the review). Peer reviewers should be independent, which is necessary for an objective and impartial evaluation of the work product. To be independent, the peer reviewer should not be associated with the generation of the specific work product, either directly by substantial contribution to its development or indirectly by significant consultation during the development of the product. In addition to being independent, peer reviewers should be impartial and free from financial COIs or other ethics issues. Disclosure of potential COIs or other ethics issues such as an appearance of a loss of impartiality—and appropriate resolution of these issues—is necessary to ensure a credible peer review.

Finally, the group of peer reviewers—whether serving on a panel or as a set of individual reviewers—should be sufficiently broad and diverse to represent fairly the scientific and technical perspectives and fields of knowledge relevant to the peer review charge. Naturally, experts whose understanding of the specific technical area(s) being evaluated are necessary; nevertheless, it also is important to include a broad enough spectrum of other related experts to consider wider dimensions of the issue(s). Although individuals who are familiar with and have a substantial reputation in the field often are called upon repeatedly to be reviewers, it is important to keep a balance by considering new individuals who bring fresh perspectives to the review of a work product. The principle is to avoid the repeated use of the same reviewer on multiple assessments unless his/her participation is essential and the expertise cannot be obtained elsewhere.

5.2.2. How Are Potential Peer Reviewers Identified?

How potential reviewers are identified depends primarily upon the peer review mechanism. Recommendations for potential peer reviewers for letter reviews or panels can be identified by a number of organizations. These include external groups, such as affected parties, special interest groups, public interest groups, environmental groups, professional societies, trade or business associations, state organizations or agencies, Native American tribes, colleges and universities, the National Research Council (NRC) and other federal agencies with an involvement in or familiarity with the issue. Recommendations for peer reviewers also may come from Agency staff, including Designated Federal Officers (DFOs) for scientific FACs—such as the Science Advisory Board (SAB), Scientific Advisory Panel (SAP) or Board of Scientific Counselors—and relevant scientific and technical experts from EPA offices.

Another method that might be used to find peer reviewers is public solicitation. The peer review plans found on the EPA Peer Review Agenda website,[28] for example, can indicate opportunities for the public to nominate peer reviewers.

If the peer review will be conducted by a contractor-managed panel, the process for identifying peer reviewers for Influential Scientific Information (ISI), including Highly Influential Scientific Assessments (HISAs), includes opportunities for the public to nominate experts and to comment on the list of candidates (See Section 4.6). In addition, the contractor may have its own pool of scientific and

---

[28] EPA. 2015. *Peer Review Agenda.* http://cfpub.epa.gov/si/si_public_pr_agenda.cfm.

technical experts for peer review. EPA may provide contractors with information on potential peer reviewers for conducting a peer review, including names if such a listing is prepared in alphabetical order. EPA should not require that the contractor select from a prepared list, nor require that the contractor receive EPA approval before selecting any given peer reviewer (sometimes known as a "subcontractor"). EPA should review the list of peer reviewers, however, for conformance to work assignment specifications (including balance of expertise) and adherence to ethics requirements before the peer reviewers are subcontracted (see Section 4.6.3). When the NAS is used to conduct a peer review, additional procedures may need to be followed (see Section 4.8).

If the peer review will be conducted using an existing EPA FAC, the DFO for the committee will take the lead for identifying peer reviewers, using a process that usually includes opportunities for public nomination and comment on candidates. An EPA office that decides to use a FAC should coordinate directly with the DFO for the FAC. For example, the SAB Staff Office publishes *Federal Register* notices to solicit names for both *ad hoc* panels and standing advisory committees. Recommendations from the EPA office requesting the peer review are considered along with public nominations and experts individually identified by the DFO. The names of candidates, along with short biographical sketches, also are posted so that the public may not only nominate, but also comment on potential advisory committee members. More information is available in the report titled *Advisory Committee Meetings and Report Development: Process for Public Involvement*, which is available from the SAB's website[29] and in Appendix F, *Guidance on Requesting a Review by the Science Advisory Board*.

In rare instances, a member of the scientific community will offer his/her services for peer review during an ongoing peer review. These offers may be at no cost or based on an expectation that reimbursement will be made. Disposition of these unsolicited offers should be handled on a case-by-case basis by the Peer Review Leader (PRL) and, as necessary, in consultation with the Peer Review Coordinator (PRC), the Office of General Counsel (OGC), Office of Regional Counsel (ORC) and appropriate Decision Makers (DMs).

For an internal peer review, the PRL will have a lead role in identifying potential EPA reviewers who have the appropriate expertise and are independent from the development of the work product. Internal reviewers should come from a different organizational unit than the one in which the work originates.

5.2.3. When Are External Peer Reviewers Preferred?

External peer reviewers are preferred for all ISI and are expected for HISAs. For some work products, such as those reviewed at various stages of product development, both internal and external peer review may be appropriate. Peer reviewers, whether external or internal, should have appropriate technical expertise, available time, and should not have been involved in the development of the work product. It should be noted that for work products categorized as HISAs, the use of internal peer reviewers is inconsistent with the guidance provided by the Office of Management and Budget's (OMB) Peer Review Bulletin. External peer reviewers could include individuals from other federal and state agencies, academic institutions and private research organizations, who possess unique or indispensable expertise.

---

[29] EPA. 2004. *Advisory Committee Meetings and Report Development: Process for Public Involvement.* http://www.epa.gov/sab/pdf/sabso_04_001.pdf. September.

5.2.4. What Should be Considered When Compiling a List of Peer Reviewers?

Usually, there is a continuum of scientific views on any issue. To the extent practicable, selected experts should include a range of technically legitimate points of view that fall along the continuum. The list of peer reviewers should include experts who are considered "mainstream" (nearer the center of the continuum), as well as those further to either side of the continuum. This will help ensure that a broad range of views will be expressed and discussed on the specific work product being reviewed, whether the objective of the peer review is to reach consensus or to provide a spectrum of views for the Agency to evaluate.

Scientific FACs are required to be balanced in terms of scientific points of view for the charge to be addressed. For example, the SAB Staff Office considers a balanced list of peer reviewers to be one characterized by inclusion of candidates who possess the necessary domains of knowledge, the relevant scientific perspectives (which, among other factors, can be influenced by work history and affiliation) and the collective breadth of experience to adequately address the charge to the peer reviewers.

For peer reviews conducted by nonfederal experts (e.g., contractors), the OMB Peer Review Bulletin directs that the evaluation of peer reviewer composition and balance be guided by NAS policies.[30]

5.2.5. Can a Foreign National Be a Peer Reviewer?

In some cases, the foremost expert in a subject area may be a citizen of another country, and the Agency may wish to obtain his or her peer review comments.

However, there are complicated legal restrictions on using foreign nationals as peer reviewers depending on whether the peer reviewer will be compensated as a Special Government Employee or an expert or consultant under 5 U.S.C. 3109, an uncompensated consultant or expert under that statute who only receives invitational travel orders, a direct contractor under a purchase order or letter contract, or a subcontractor to a prime contractor. If this issue comes up, EPA's PRL should consult with OGC and the Office of Human Resources (OHR).

5.2.6. Are There Other Constraints to Selecting Peer Reviewers?

5.2.6.1. Timing

Sometimes the schedule for a peer review is accelerated because of a court-ordered deadline or other time-sensitive requirement. In these cases, there may be constraints in selecting peer reviewers and conducting a peer review in a timely manner. Processes should be developed for identifying and using a small number of peer reviewers to ensure that quick, effective peer review can be included for even the most rapidly moving products.

5.2.6.2. Confidential Business Information (CBI)

Another possible constraint involves CBI. There are different definitions and types of CBI, depending on the statute that governs the action. To evaluate certain Agency-generated studies properly, some peer reviewers may need access to CBI. If the reviewers are federal employees or contractors/subcontractors with CBI clearance, the Agency does release CBI to them. Therefore, whenever contemplating the use

---

[30] National Academy of Sciences. 2003. *Policy on Committee Composition and Balance and Conflicts of Interest for Committees Used in the Development of Reports.* Washington, D.C.: The National Academies Press. http://www.nationalacademies.org/coi/bi-coi_form-0.pdf.

of outside peer reviewers, Agency staff should determine whether the reviewers will need access to CBI. If they do not have CBI clearance, OGC should be consulted on whether it is practical to obtain the consent of affected CBI submitters to disclose the information to peer reviewers.

5.2.6.3. Lobbyists

In accordance with a 2010 Presidential Memorandum and OMB revised guidance to implement the policy (79 FR 47482, August 13, 2014), no RGE or SGE member of a FAC appointed to serve in an individual, expert capacity may be a federally registered lobbyist. This prohibition does not apply to FAC members who are appointed to serve in a representative capacity on behalf of an interest group or constituency.

5.2.7. Can Someone Who Provided Peer Input Become an Independent Peer Reviewer for the Same Work Product Later in the Process?

Generally, the answer is no, because that expert is no longer independent but rather is a contributor to the work product. There may be special circumstances under which the expertise is so narrow that another peer reviewer is not available. The PRL normally will be responsible for making this determination and documenting the decision in the peer review record.

5.2.8. Can a Peer Reviewer Be Used to Review the Same Product More Than Once or to Review Multiple Products?

There is no prohibition against using the same peer reviewer more than once on the same product or for multiple products of the same EPA office. It is preferable, however, to use different individuals each time the product is sent back for peer review to provide a broader perspective. It is particularly important to rotate peer reviewers across the pool of qualified reviewers in the case of multiple HISAs. In the case of sequential reviews of one product, it can be beneficial to seek review from the same individuals where the review is focused on revisions made to address the peer reviewers' comments. Even in such cases, it may be helpful to include reviewers who were not involved in the previous review of the product to ensure that the product gets a fresh look.

When using a contractor to provide peer review services, it should be considered that contractors may have a "pool" of reviewers that they use regularly. If the same peer reviewers are used repeatedly, they may lose their impartiality (or the appearance of impartiality) relative to the work product(s). In addition, there may be competition or directed subcontracting issues when seeking subsequent reviews of a work product by the same peer reviewers if they were appointed under a contract mechanism. If there is a possibility that the same peer reviewers may be needed to conduct subsequent peer reviews of work products, the Contracting Officer (CO) must be informed when the contract for the initial review is being planned. In most cases, competition or other contracting issues that might complicate or preclude the use of the same peer reviewers for subsequent reviews of the same work product can be addressed with a properly drafted Statement of Work (SOW) and appropriate contract clauses.

When seeking the same peer reviewers for additional rounds of peer review, the peer reviewers should be reevaluated for independence, COI and appearance of a loss of impartiality before they serve as a repeat panel member. The appropriate peer review COI form should be used to identify any potential COI that may have arisen since completion of the previous round of peer review (see Appendix J).

5.2.9. If State or Tribal Employees Are Used as Peer Reviewers, Can EPA Pay Them for This Service?

In some cases, this may be possible. However, the PRL should ensure that the state or tribal employee has received the necessary approvals since providing a state or tribal employee with compensation as an expert or consultant under 5 U.S.C. 3109 or as a direct EPA contractor may conflict with the state or tribe's ethics or personnel laws or policies. Similarly, the PRL should ensure that peer review prime contractors verify that state and tribal employees may work as subcontracted consultants before hiring them.

EPA would be able to pay travel expenses under a 5 U.S.C. 5703 invitational travel order since the "consult with or otherwise provide a direct service to EPA" requirement would be met and most states and tribes allow their employees to accept invitational travel orders. If the state or tribal employee will not be paid for their peer review services, the letter or email inviting them to the peer review meeting should make it clear that EPA is only providing travel support; the letter or email must also clearly indicate that the state or tribal employee will provide peer review services to EPA without compensation and that the state or tribal employee will make no future claim for compensation for the peer review services.

Please note that because peer reviewers provide services for EPA's direct use or benefit, states and tribes may not charge federal grants/cooperative agreements for their employees' time or travel while working on a peer review due to the Federal Grant and Cooperative Agreement Act.

If the state or tribal expert is not being paid for his/her peer review services, or reimbursed for travel expenses, he/she must sign an agreement stating that he/she does not expect payment. See Section 4.6.7 for information on gratuitous services agreements.

5.2.10. Can the Identity of Peer Reviewers Be Kept Anonymous by EPA?

No, the identity of peer reviewers cannot be kept anonymous by EPA. However, the attribution of specific comments to any given peer reviewer is not necessary. Peer reviewers should be informed in advance of EPA plans for releasing their names and credentials, as well as the extent of attribution of comments to specific reviewers. If a peer reviewer requests anonymity at the outset of the peer review, the PRL should inform the peer reviewer that there is no guarantee of anonymity. Although this may be a deterrent to possible peer reviewers, EPA is committed to working with the fullest possible transparency to the public (except where statutorily constrained, such as with CBI).

The reviewers' names and affiliations may be made available to the public before the review begins depending on the peer review process used. For all ISI and HISAs, the names and affiliations of peer reviewers should be listed in the peer review reports. Release of any reviewer information retrieved by a personal identifier must be performed in accordance with the Privacy Act of 1974 (5 U.S.C. § 552(a), as amended), as interpreted in OMB implementing guidance, 40 *Fed. Reg.* 28,948 (Jul. 9, 1975).

For other types of peer reviews that do not qualify as ISI or HISA, such as the peer review of extramural grant applications, reviewer names can be held in anonymity to the public, unless, in some circumstances, they are requested under the Freedom of Information Act (FOIA). If a request for peer review documents is received under the FOIA, the requestor may be able to view any comments attributed to specific reviewers.

5.3. Ensuring a Credible Peer Review Process – Ethics Considerations

5.3.1. What Are the Relevant Ethical Standards for Different Categories of Peer Reviewers?

To ensure a credible peer review process, PRLs must ensure that the appropriate and relevant ethical standards are applied to each of the peer review mechanisms. When a peer review panel is used, ethical standards must be adhered to not only during the panel formation process, but also during and after the peer review itself has been completed. These ethical standards are embodied in the various laws, implementing regulations and other requirements that apply to peer reviewers who are RGEs, SGEs, contractors and those who are selected by outside organizations (e.g., the NAS) (see Table 2). For peer reviews conducted by outside organizations, the PRL should be thoroughly familiar with the ethics

Table 2. The Applicable Rules for Conflict of Interest and Impartiality of Peer Reviewers Depends on the Status of the Peer Reviewers

Peer Review Approach	Status of Peer Reviewer	Applicable Conflict of Interest/Ethics Rules and Policies	Handbook Section
Internal	RGE	18 U.S.C. § 201, 203, 205, 207, 208 and 209; Standards of Ethical Conduct in the Executive Branch	5.3.3, 5.3.6–5.3.9
External: Publication in Refereed Journal	Independent experts selected by the journal	COI/ethics rules of the journal	n/a
External: Letter Reviews	Contractor, Subcontractor	FAR, EPA Acquisition Regulations, contract terms and conditions	4.6
	Gratuitous Services Peer Reviewer	Gratuitous services agreement terms	4.6.7
External: Contractor Panel	Contractor, Subcontractor	EPA process on contractor-managed peer review panels for review of ISI/HISAs, FAR, EPA Acquisition Regulations, contract terms and conditions	4.6
External: FACA Panel	SGE, non-EPA RGE	18 U.S.C. § 201, 203, 205, 207, 208 and 209; Standards of Ethical Conduct in the Executive Branch; Presidential Memorandum *Lobbyists on Agency Boards and Commissions* (June 18, 2010)	5.2.6, 5.3.3–5.3.9
External: NAS/NRC	Independent experts selected by the NAS/NRC	NAS *Policy on Committee Composition and Balance and Conflicts of Interest*[31]	4.8

policies and requirements of the organization conducting the review. For example, if a peer review is to be conducted by the NAS, the PRL would need to be familiar with the NAS *Policy on Committee*

---

[31] National Academy of Sciences. 2003. *Policy on Committee Composition and Balance and Conflicts of Interest for Committees Used in the Development of Reports*. Washington, D.C.: The National Academies Press, 2003. http://www.nationalacademies.org/coi/bi-coi_form-0.pdf.

Composition and Balance and Conflicts of Interest.[32] For work products that are peer reviewed through publication as journal articles, the ethics standards and processes are set by the specific journal.

Specific regulations have set forth ethics considerations for contractor personnel and government employees. An overview of these requirements as they relate to peer review is provided in the sections that follow. This chapter focuses primarily on ethics issues such as potential COIs or an appearance of a loss of impartiality for government employees, including SGEs. See Chapter 4 for a discussion of COI for peer reviews obtained by contract.

5.3.2. How Are Ethics Issues Evaluated for Peer Reviewers Under Contractor-Managed Peer Reviews?

As noted in Section 4.6, contracts to obtain peer review services should include COI clauses, and COI requirements should be included in the SOW. Information necessary to ensure that peer reviewers are free from ethics issues such as potential COIs or an appearance of a loss of impartiality is collected by the contractor managing the peer review using a confidential peer review COI questionnaire pursuant to the peer review "Conflict of Interest Evaluation for Task Orders/Work Assignments" clause and other COI requirements included in the contract. The PRL should work with the CO to assist the contractor with identifying actual or potential COI that might impair the objectivity of peer reviewers. In the case of a contractor-managed panel review of ISI or HISAs, oversight by the CO of the contractor's identification and proposed resolution of COI issues should include consultation with the EPA Science Advisor or his or her designee.

5.3.3. How Are Ethics Issues Evaluated for Peer Reviewers Who Are Government Employees?

The PRL (and appropriate EPA ethics officials) will typically consider five COI statutes: 18 U.S.C. §§ 203, 205, 207, 208 and 209. In addition to these COI statutes, all government employees, including SGEs, must adhere to the Standards of Ethical Conduct for Employees of the Executive Branch (5 C.F.R. 2635). Although responsibility for compliance rests with the individual government employee, PRLs and appropriate ethics officials must work together to ensure that all applicable ethics laws and implementing regulations are followed when government employees are peer reviewers (e.g., internal peer review by EPA experts, external peer review by SGE or non-EPA RGEs).

5.3.4. What Constitutes a Conflict of Interest for a Special Government Employee on a Federal Advisory Committee?

SGEs typically have outside (i.e., non-EPA) employment as well as other financial interests, which may potentially present COI issues under 18 U.S.C. § 208 (a criminal COI statute). According to this law, government employees (including SGEs) are prohibited from participating personally and substantially in any particular matter that has a direct and predictable effect on their own financial interests or the financial interests of others whose interests are imputed to them. For a COI to be present, all of these elements must be present. If an element is missing, there is no COI.

---

[32] National Academy of Sciences. 2003. *Policy on Committee Composition and Balance and Conflicts of Interest for Committees Used in the Development of Reports.* Washington, D.C.: The National Academies Press. http://www.nationalacademies.org/coi/bi-coi_form-0.pdf.

For example, ownership of stock is not a COI absent personal and substantial participation by an SGE in a particular matter that will have a direct and predictable effect on this interest.

To apply ethics regulations to FAC members properly, it is important to know whether the charge to a committee is a "matter," a "particular matter of general applicability" or a "particular matter concerning specific parties." A matter is something that is directed to the interests of a large and diverse group of persons. A particular matter of general applicability is focused on the interests of a discrete and identifiable class of persons (e.g., a certain industrial sector). A particular matter concerning specific parties is focused on the legal rights of parties or transactions (e.g., grants, contracts, investigations, litigation). When a charge is not a particular matter, then 18 U.S.C. § 208 does not apply, and a COI cannot arise. Furthermore, particular matters of general applicability and particular matters concerning specific parties are treated somewhat differently in the ethics regulations, as explained in Section 5.3.7.

5.3.5. Can a Recipient of EPA Contracts or Grants Be a Peer Reviewer?

EPA frequently issues contracts to develop scientific and technical work products for the direct benefit of or use by the Agency. Contractors who help develop those work products are not independent of the work product and cannot serve as peer reviewers of the same work product. Even if a contractor is not involved in the development of a work product being reviewed, the nature and extent of his or her contractual relationships with the Agency or with the EPA office sponsoring a peer review should be considered when selecting reviewers to ensure that the contractor is sufficiently independent from the Agency or EPA office as a general matter.

EPA also provides grant money through competitive processes to further the investigation of science matters it believes would benefit its mission. As noted in the OMB Bulletin, when a scientist is awarded an EPA research grant through an investigator-initiated, peer-reviewed competition, there generally should be no question as to that scientist's ability to offer independent scientific advice to the Agency on other projects. Those grantees are independent of Agency direction, and can serve as peer reviewers for scientific or technical work products (or portions thereof) that are not solely a product of their own research conducted under the Agency grant. For example, a grantee may review a work product that synthesizes a body of literature, such as an integrated science assessment, that happens to incorporate agency funded work conducted by the grantee. The grantee must, however, still be free from financial COI or the appearance of a loss of impartiality (see Sections 5.3.4 and 5.3.7).

PRLs may question whether experts who currently receive funding from EPA (e.g., grants, contracts, assistance agreements) have, by definition, an inherent financial COI and therefore cannot be peer reviewers. If an expert previously received funding, but does not currently, then there is no financial COI. If an expert is currently receiving funding through an EPA grant, the PRL should examine how the grant was awarded. If EPA awarded the grant through a competitive, peer-reviewed process, then the Agency's ability or potential to influence the expert's research is limited. Consequently, there is little likelihood that the expert's ability to offer scientific advice is subject to any financial COI.

Alternately, if an expert has an existing consulting or contractual arrangement with the Agency, then the expert is beholden directly to EPA on closely-related matters. Consequently, this situation presents a greater concern about appearance of a financial COI.

5.3.6. Are There Any Exemptions or Remedies from a Conflict of Interest for Regular and Special Government Employees?

5.3.6.1. Conflict of Interest Exemptions for Special Government Employees

SGEs serving on FACs specifically are exempted by regulation from certain provisions of the financial COI statute (18 U.S.C. § 208). An exemption (5 C.F.R. 2640.203(g)) permits SGEs serving on FACs to participate in particular matters of <u>general applicability</u> when the disqualifying interest arises from the SGE's nonfederal employment or prospective employment. Whenever there are questions about COIs, the PRL should contact the appropriate Deputy Ethics Official (DEO) and/or OGC/Ethics, who in turn may consult with the U.S. Office of Government Ethics (OGE) for assistance (http://intranet.epa.gov/ogc/ethics/deos.htm).

It is important to note that the exemption does not extend to the SGE's financial holdings or consultancies. Furthermore, this exemption is subject to several limitations:

- The matter cannot have a "special or distinct" effect on either the SGE or the SGE's nonfederal employer, other than as part of a class;

- The exemption does not cover interest arising from ownership of stock in the employer; and

- The nonfederal employment must involve an actual employer-employee relationship, as opposed to an independent contractor relationship.

5.3.6.2. Conflict of Interest Remedies

COI may be remedied through nonparticipation in the matter (also known as "recusal" or "disqualification"), divestiture from the disqualifying interest, or the granting of a waiver pursuant to provisions under 18 U.S.C. § 208(b). Whenever there are questions about COIs, the PRL should contact the appropriate DEO and/or OGC, who in turn may consult with the OGE for assistance (http://intranet.epa.gov/ogc/ethics/deos.htm).

- **Nonparticipation.** COI may be remedied by nonparticipation in a particular matter. Nonparticipation means that the employee does not participate personally and substantially in the particular matter. When a panel considers more than one particular matter, it is possible for an employee to recuse himself/herself from only those particular matters for which he or she has a COI.

- **Divestiture.** COIs may be remedied by divestiture from the disqualifying interest to below certain *de minimis* exemption levels. These exemption levels vary depending upon the type of particular matter being considered (see 5 C.F.R. § 2640.201 for more information on exemptions available for RGEs and SGEs). When divestiture from a disqualifying interest is sought as a remedy for a potential COI, it should be noted that SGEs (as opposed to RGEs) are not eligible for a "certificate of divestiture" that allows for deferral of capital gains in the divested asset.

- **Statutory Waivers from COI.** An SGE who serves on a FAC may seek a waiver from OGC to participate under the provisions of 18 U.S.C. § 208(b)(3). Only the EPA's Designated Agency Ethics Official (DAEO) can grant such a waiver, and only if he/she certifies in writing (in consultation with OGE) that the need for the SGE's services outweighs the potential for a COI posed by the financial interest involved. SGEs not serving on FAC (and all RGEs) may be considered for waivers only in accordance with the more restrictive standard under 18 U.S.C. § 208(b)(1), which requires a determination by the DAEO that the financial interest is not so substantial as to be deemed likely to affect the integrity of the employee's services. Further guidance on waivers may be found in OGE DO-07-006 ("Waivers under 18 U.S.C. § 208").[33]

5.3.7. What Is an Appearance of a Loss of Impartiality for Regular and Special Government Employees?

When forming peer review panels with RGEs/SGEs, another common ethics issue that may arise is an "appearance of a loss of impartiality" as defined by 5 C.F.R. Part 2635, Subpart E. PRLs must be alert not only to COI issues (which tend to be easier to recognize), but also to "appearance" issues, which can be more subtle. Unlike COI issues, appearance issues do not violate any criminal statute. An appearance of a loss of impartiality may occur when an employee's participation in a particular matter involving specific parties (e.g., a contract, an enforcement action) might cause a reasonable individual with knowledge of the relevant facts to question that employee's impartiality. Appearance issues arise if the peer review activity is likely to have a direct and predictable effect on the financial interests of a member of a peer reviewer's household, or if the peer reviewer has a "covered relationship" (as defined in 5 C.F.R. 2635.502(b)) with someone who is (or represents) a specific party or parties involved in the matter.

For example, if a member of an employee's household (e.g., a relative with whom the employee has a close personal relationship) has a contract with a company to conduct all of the marketing for a pesticide that has a pending registration before the Agency, and the Agency is convening a peer review panel to evaluate a study that will be pivotal in determining whether to grant the registration (a specific party matter), then a reasonable individual would question the employee's ability to participate impartially in the peer review.

If an employee's participation in a peer review would cause a reasonable individual to question the employee's impartiality, the appropriate DEO in the organization conducting the peer review may authorize the employee to participate in the review based on a determination, made in light of all relevant circumstances, that the interest of the government in the employee's participation outweighs the concern that a reasonable individual might question the integrity of the Agency's programs and operations. For discussion of factors that should be considered when deciding whether to authorize participation, see 5 C.F.R. 2635.502(d). After considering these factors, the appropriate DEO may decide to authorize the employee's participation or, conversely, to prohibit it. Regardless of the outcome, OGC strongly recommends that the DEO issue a written determination that documents the final decision.

5.3.8. How Should Peer Review Leaders Address Ethics Issues for Regular and Special Government Employees during Peer Reviewer Selection?

The peer reviewer selection process is the step in the peer review process when the PRL is most likely to initially encounter ethics issues such as potential COIs or an appearance of a loss of impartiality. To evaluate potential issues, financial disclosure forms are obtained and evaluated by the appropriate ethics

---

[33] Cusick, Robert I., Director, Office of Government Ethics. 2007. Memorandum to Designated Agency Ethics Officials. *Waivers Under 18 U.S.C. § 208.* DO-07-006. http://www.oge.gov/OGE-Advisories/Legal-Advisories/DO-07-006--Waivers-under-18-U-S-C--%C2%A7%C2%A7-208(b)(1)-and-(b)(3)/. February 23.

official (usually the responsible DEO in the EPA office where the peer review takes place). For RGEs, either OGE Form 450 (Confidential Financial Disclosure Form) or OGE Form 278 (Public Financial Disclosure Form) is collected and evaluated. For SGEs, EPA Form 3110-48 (Confidential Financial Disclosure Form for Special Government Employees Serving on Federal Advisory Committees at the U.S. Environmental Protection Agency) is typically filed with the DFO's own DEO. In rare instances, however, an SGE or an RGE may be required to file the OGE-278 (Public Financial Disclosure Report). This report is filed with OGC, along with any necessary OGE-278T (Periodic Financial Transaction) forms. In all instances, financial disclosure forms are filed and reviewed, both annually (with some exceptions) and prior to any new matter.

Before finalizing the selection of reviewers, the PRL should ascertain whether each potential peer reviewer's involvement in certain activities could pose an ethics issue such as potential COIs or an appearance of a loss of impartiality. Each matter should be treated on a case-by-case basis and can involve a number of factors. Employment and professional affiliations of the participants, as well as their financial interests, should be considered. Some actions that should be taken in evaluating ethics issues include, but are not limited to, the following:

- Discussing ethics issues with each participant before the review process takes place.

- Disclosing publicly at the beginning of meetings any previous involvement with the matter.

- Obtaining appropriate and up-to-date financial disclosure forms.

- Collecting additional information through public comment and other appropriate means.

5.3.9. What Other Ethics Issues Might Arise for Regular and Special Government Employees During or After a Peer Review?

Peer reviewers who are government employees, including SGEs, are subject to ethics requirements in addition to those regarding COI under 18 U.S.C. § 208 or an appearance of a loss of impartiality during panel operation and even after a panel has completed its work. Therefore, it is prudent to inform SGEs both prior to and during their service that ethics requirements such as postemployment restrictions may apply to them, dependent on the type of particular matter they worked on as well as the level of compensation they received during the time of service. These issues are discussed in-depth in EPA Ethics Advisory 08-02; some of the more typical restrictions are summarized below:

- **Representational Activities Directed Toward the United States.** Two companion ethics laws (18 U.S.C. §§ 203, 205) prohibit an employee from representing outside organizations and individuals on any particular matter in which the United States is a party or has a direct and substantial interest, before any department, Agency or other specified entity, whether for compensation or not. For SGEs, these statutes apply only with respect to particular matters involving specific parties (e.g., contracts, grants, enforcement actions), and their application depends on the number of days that the SGE worked for the federal government in the preceding 365-day period.

- **Compensation for Teaching, Speaking or Writing on Matters Related to Official Duties.** In certain cases, SGEs are prohibited from receiving outside compensation for teaching, speaking or writing when the activity is undertaken as part of their official EPA duties. SGEs also are subject to the criminal bribery and illegal gratuity statute, which prohibits, under certain circumstances, the receipt of anything of value in connection with official acts.

- **Hatch Act Political Activity Restrictions (5 U.S.C. §§ 7321 – 7328).** The Hatch Act places some restrictions on federal government employees, including SGEs, when they engage in partisan political activity. During the time that SGEs are actually performing government business, they are prohibited from any fundraising for any partisan political group, candidate or campaign. They cannot engage in partisan political activity while on duty or while using a government vehicle, or in any room or building used for government business, and cannot use their SGA affiliation in connection with such political activity.

- **Seeking Other Employment (5 C.F.R. Part 2635, Subpart F).** SGEs may be interested in seeking other nonfederal employment while serving as government employees. SGEs may not participate in any particular matter that directly and predictably affects the financial interest of any individual or organization with whom/which they are seeking future employment, contracts or consultancies unless authorized by the appropriate DEO (who is required to consult with OGC) or, if the COI restriction at 18 U.S.C. § 208(a) applies, they have been granted a waiver under 18 U.S.C. § 208(b)(1). Such waivers are rarely, if ever, granted by OGC. It also is noted that under a provision of the Stop Trading on Congressional Knowledge Act, SGEs filing a public financial disclosure form (OGE-278) must report any postemployment negotiations to OGC within 3 business days. More information on Stop Trading on Congressional Knowledge Act requirements (including periodic financial transaction reporting) is available from OGC.

- **Service as an Expert Witness (5 C.F.R. § 2635.805).** On occasion, the PRL may find that an SGE wants to serve as an expert witness for an outside organization. Government employees cannot serve (other than on behalf of the United States) as an expert witness before a court or agency of the United States in which the United States is a party or has a direct and substantial interest, unless authorized by the agency's DAEO. For SGEs, the number of days of service affects the ban on serving as an expert witness.

- **Postemployment Restrictions.** Former employees (including SGEs) are prohibited by federal law (18 U.S.C. § 207) from making representations on behalf of another back to the federal government with the intent to influence a federal official with respect to any particular matter involving specific parties in which the United States is a party or has a direct and substantial interest. Thus, for example, a former SGE who served on a FAC cannot represent an outside individual or organization back to the federal government (not just this Agency) concerning the same specific party matter that was the subject of the FAC. SGEs working more than 60 days in any 365-day period who file an OGE-278 public financial disclosure report are subject to a 1-year "cooling off" period and cannot make representations on behalf of another back to EPA with the intent to influence any official action regardless of whether the SGE participated in it personally and substantially and regardless of whether the matter involves specific parties or not.

6. Conducting and Completing the Peer Review

6.1. Overview

For a peer review to be successful, peer reviewers should receive several documents at the beginning of the process (Figure 7). The specific documentation to be provided is based on the type and mechanism of peer review to be conducted, as discussed in Chapter 4. In each case, peer reviewers should be given what is necessary to complete their task; however, they should not be overburdened, with excess material. Needed documentation includes, but is not limited to, the work product to be reviewed, a clear charge and logistical details.

6.2. The Peer Review Charge and Instructions to Peer Reviewers

6.2.1. What Is a Charge?

A charge is a set of focused questions that identifies the scientific and technical issues on which the Agency would like feedback and invites suggestions for improving the document as a whole. The charge should be developed prior to the selection of the peer reviewers to ensure availability of appropriate scientific and technical expertise and skills for reviewing the specific work product. Preparing a good charge is time well-spent, as the charge is crucial for an effective peer review. A good charge will direct the reviewers to give advice on issues relevant to the Agency and will lead to a greater understanding of the reviewer's reasoning, which is pivotal to the Agency's ability to address the reviewers' concerns and to craft specific improvements to the work product (see Appendix H).

Generally, the charge to peer reviewers includes two types of questions. The first type identifies specific technical and scientific issues about which the Agency would like feedback. These focused charge questions should be explicit enough to encourage constructive comments, but not so narrow that they

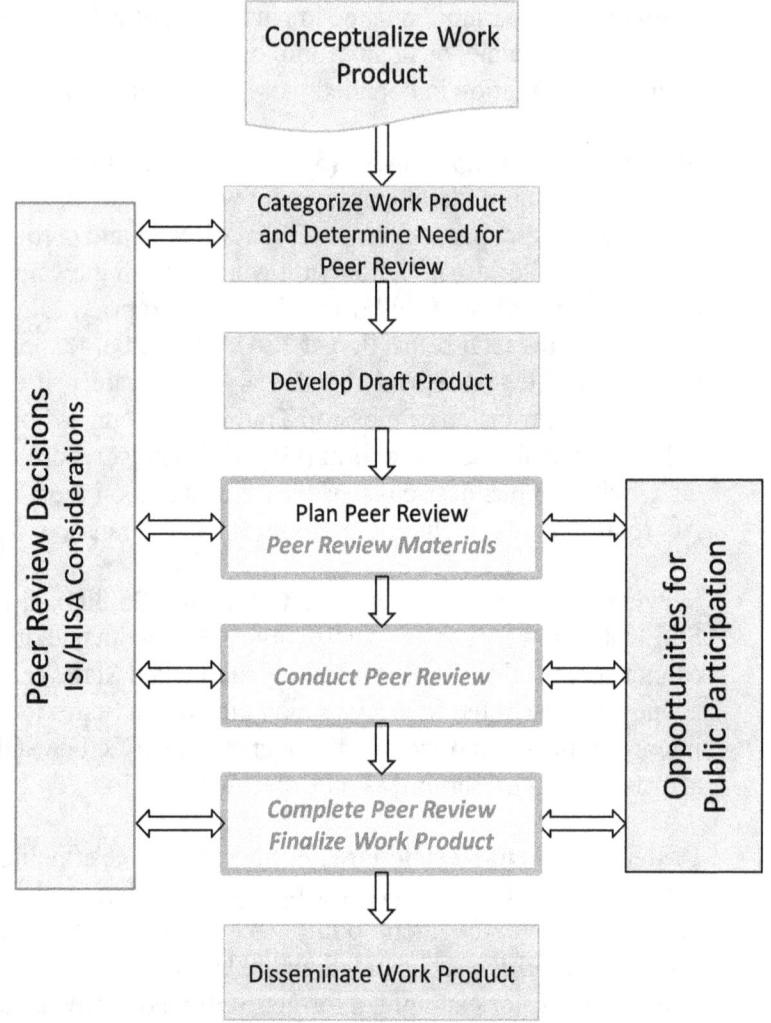

Figure 7. The Peer Review Process: Conduct and Complete Peer Review

preclude or limit informative responses that the reviewer may consider important to provide. The second type of question typically invites a broad evaluation of the overall work product. It is important to remember, however, that the peer review is not conducted for the

Time is well-spent preparing a good charge, as the charge is crucial for an effective peer review.

purpose of evaluating a potential Agency action, decision or policy. Reviewers should not be asked to advise the Agency on policy.

6.2.2. What Are the Essential Elements of a Charge?

A well-prepared charge includes:

- A concise overview or introduction describing the work product, its development and its intended use.

- Issues to be addressed and areas of concern or specific advice sought (in the form of charge questions), such as:

  o The soundness of the method(s) used or proposed.

  o The scientific support for the assumptions employed.

  o The identification of scientific uncertainties and the potential implications of those uncertainties for the stated conclusions and for influential scientific information (ISI) and highly influential scientific assessments (HISAs), that scientific uncertainties are clearly identified and characterized.

  o Recommendations for research that would reduce key uncertainties.

  o The sensitivity of the results to alternative assumptions (i.e., sensitivity analysis).

  o The comprehensiveness and utility of the literature reviewed.

In addition, a request may be made for the reviewers to raise issues that might not have been considered by the authors in their charge questions. Examples of peer review charges that have been used successfully by the Agency and cover a variety of issues are provided in Appendix H.

6.2.3. Can the Public, Including Stakeholders, Provide Input to the Charge to the Peer Reviewers?

Yes, depending on the type of peer review (e.g., letter review versus panel review), availability of a public version of the draft work product, resources and other factors (e.g., timing), EPA may obtain public input regarding the charge to the peer reviewers. (Note that this would require releasing the draft work product.) The Agency makes the final determination, however, on what elements to include in the charge to ensure that it meets the EPA's needs for the peer review. The following considerations should be taken into account:

- The Agency can obtain public input regarding the charge via a notice on the EPA Web page or through a *Federal Register* notice.

- If stakeholder input is sought, interested parties should be included to the extent feasible given statutory, regulatory, budgetary and/or time constraints. Input should not be limited to a single stakeholder or to one side of a controversial issue (e.g., a responsible party or environmental group).

- If a group is formed to help develop the charge, care should be taken to ensure that the group does not become subject to the requirements of the Federal Advisory Committee Act (FACA).

- If an annotated outline or draft of the work product can be shared with the public, this will facilitate public input on the charge.

6.2.4. Who Writes the Charge When the Agency Hires a Contractor to Conduct the Peer Review?

In general, if EPA uses a contractor to manage the peer review, EPA should allow the contractor independence in conducting it. However, to ensure that the peer review meets the EPA's needs, EPA personnel are responsible for providing the list of charge questions to the contractor managing the peer review for distribution to the peer reviewers. If the charge questions are known prior to the issuance of a solicitation for a contract, or prior to the issuance of a tasking document under an awarded contract, the Contracting Officer's Representative (COR) can incorporate the charge questions directly into the Statement of Work (SOW).

EPA may task the contractor with providing advice and assistance in developing some elements of the charge, such as the overview or introduction. In such cases, however, EPA personnel are still responsible for providing the contractor with the list of questions to be included. Whenever the contractor assists EPA in developing the charge, EPA must review and comment on a draft of the charge and approve any revisions to it.

The EPA cannot submit the charge or the charge questions directly to the peer reviewers when the review is being managed by a contractor. Rather, the contractor is responsible for submitting the charge to the reviewers along with other review materials.

For details on peer reviews conducted by the National Academy of Sciences (NAS), see Section 2.3.7.

6.2.5. What Additional Instructions and Information Does the Agency Give Peer Reviewers, including Preparation of a Peer Review Report?

6.2.5.1. General Instructions

The Peer Review Leader (PRL), or contractor (in the case of a contractor-managed peer review), provides the peer reviewers with a peer review package that includes the draft work product, charge and other pertinent material. For HISAs, the peer reviewers should be given background information about studies or models that support the key findings and conclusions of the Agency's draft assessment.

The Agency (or the contractor managing the peer review) should instruct peer reviewers as follows:

- Peer reviewers are to advise the Agency or contractor (in the case of a contractor-managed peer review) of ethics issues, including actual or potential organizational or personal conflicts of interest (COI) or other matters that would create the appearance of a loss of impartiality, guidance on which is provided in Section 5.3.

- Peer reviewers are to provide written comments (if a letter review) or a peer review report (if a panel review) (see Table 1) responsive to the charge in a specified format by a specified deadline.

- Peer reviewers are to comply with requests for confidentiality, if any, regarding the release of draft Agency products, positions or other materials provided to the reviewer. Unless the peer review is being conducted by a federal advisory committee (FAC), material provided as part of the review should be kept confidential and should not be discussed outside the designated panel discussion times or shared outside the panel.

- From the time they accept the invitation to review the work product, peer reviewers should avoid interactions—including with Agency representatives or members of the interested public—that might create a perception of COI regarding the work product under review.

- Members of peer review panels, either Agency-led or contractor-managed, should immediately inform the PRL or contractor if they are contacted regarding the peer review or work product by anyone other than another panel member. The contractor will immediately inform the COR of any reports by panel members of pre-meeting contacts to guard against inappropriate influence from outside the panel.

Finally, peer reviewers should receive logistical details regarding the review, such as:

- The due date for comments (for a letter review) or peer review report (for a panel review).

- Times and locations of meetings, if applicable.

- The planned extent of disclosure of reviewers' names and attribution of comments.

- The point of contact. When reviewers are selected by a peer review contractor, the point of contact should be an employee of the contractor, not an employee of the Agency.

- Type of peer review report and mode of delivery.

6.2.5.2. Further Instructions to Peer Reviewers of ISI and HISAs

For work products categorized as ISI or a HISA, peer reviewers should be instructed to prepare comments or a peer review report that describes the nature of their review, findings and conclusions. The peer review report either should: (1) include a verbatim copy of the individual reviewers' comments (with or without specific attributions); or (2) represent the views of the group as a whole, including any disparate and dissenting views, for contractor-managed panel peer reviews or FACs (although attribution of comments to names is not necessary). The peer review report should include the names of the reviewers and their organizational affiliations. For HISAs, the peer review report also should include a copy of the charge to the reviewers and a short paragraph on the credentials and relevant experience of each reviewer. The format and level of detail should be consistent across reviewers. Peer reviewers' written comments may be made publicly available via the EPA website, and peer reviewers should be informed of this possibility.

6.2.6. How May EPA Interact With External Peer Reviewers During the Review?

6.2.6.1. When EPA Conducts an External Peer Review

The PRL normally has administrative contacts with the reviewers during the development and conduct of the peer review. In some cases (e.g., a Science Advisory Board [SAB] peer review), peer reviewers also may receive a briefing from Agency personnel on the product to be peer reviewed. For external peer reviews conducted by FAC panels, the Designated Federal Officer (DFO) serves as the liaison between the peer reviewers and the EPA office requesting the review, as well as between the peer reviewers and members of the public. Otherwise, the PRL and other EPA office staff should not contact the reviewers during the course of the review. Such contact can lead to perceived inappropriate direction that could compromise the independence of the review.

6.2.6.2. When a Contractor Conducts an External Peer Review

If peer review is conducted via a contract under which the contractor manages the peer review(s), EPA should limit direct contact to the prime contractor's designated representative and should not have general contact with, or provide direction to, the contractor's staff or peer reviewers (subcontractors). Contact with the contractor should only be through the Contracting Officer (CO) or COR.

6.2.7. When May the Public Provide Comment During the Peer Review?

Whenever feasible, EPA offices should make drafts of work products categorized as ISI and HISAs available to the public for comment, as well as a draft peer review charge, at the same time they are submitted for peer review. For work products that are not influential, a public comment period still may be beneficial. Accepting public comments before peer review has two benefits: (1) the Agency can consider public comments on the scope of the charge before the selection of peer reviewers so that appropriate expertise is included to address all charge questions; and (2) the Agency's public comment process is kept distinct from the peer review panel's comment process. When employing a public comment process prior to the peer review, EPA offices should provide peer reviewers with access to public comments that address significant scientific or technical issues whenever practical.

When peer review of a HISA is conducted by a panel (either contractor-managed or by a FAC), the process should include a public meeting, whenever feasible and appropriate. During this public meeting, interested members of the public can make oral presentations on scientific issues relevant to the topic under review. To ensure that public participation does not delay activities unduly, EPA offices should specify time limits for public participation throughout the peer review process. It is recommended that the EPA Docket (available at http://www.regulations.gov) be used as the repository for public comments. To establish a docket, see http://intranet.epa.gov/fdmsinfo.

6.3. Responding to Peer Review Comments

6.3.1. How Does the Agency Evaluate and Incorporate Peer Reviewers' Comments?

Although the Agency is not obligated to take all recommendations provided by peer reviewers, all reviewer comments should be considered and incorporated where relevant and appropriate. For letter and panel peer reviews, the Agency evaluates the comments and prepares the response. The PRL and/or Project Manager (PM) should evaluate and analyze all peer review comments and recommendations carefully. As discussed earlier, a carefully crafted charge to the peer reviewers simplifies the organization and analysis of comments. The appropriateness and objectivity of the comments should be evaluated. Analyses may include consultation with other personnel within EPA.

The PRL and/or PM should brief the Decision Maker (DM), as appropriate, as well as all appropriate managers in the their chain of command, on the peer review comments, and should provide a proposal on how to address the peer review comments. The PRL and/or PM should identify clearly for the DM any key peer review comments, including significant comments that will not be accepted and why, as well as any controversial comments that need resolving. Comments that may lead to allocation of additional resources or a revised schedule for the completion of the work product are particularly important and should be evaluated in consultation with management.

Adequate documentation is needed to show whether comments were accepted or rejected. The documentation may be brief, but it should address all relevant and appropriate comments. The peer review record should contain a document describing the Agency's response to the peer review comments.

When peer review is conducted through a journal, the individual authors of the article evaluate and respond to the peer review comments.

6.3.2. How Does the Agency Address Comments from Peer Review Reports?

Reviewers of work products categorized as ISI and HISAs are asked to produce and submit the peer review report describing the nature of their review, and their findings and conclusions. EPA offices are expected to make these reports publicly available to implement the provisions of the Office of Management and Budget's (OMB) Peer Review Bulletin. The EPA posts or provides a link to the peer review reports, along with all materials related to the peer review, on its publicly accessible EPA Peer Review Agenda website (Science Inventory [SI] website [http://epa.gov/si/]).

The credibility of the final influential work product is likely to be enhanced if the public understands how the Agency addressed the specific concerns raised by the peer reviewers. Therefore, for HISAs, EPA offices should prepare a written response to comments in the peer review report explaining (1) the Agency's agreement or disagreement with the views expressed in the report; (2) the actions that have been or will be taken to respond to the report; and (3) the reasons that the EPA office believes those actions satisfy any key concerns or recommendations in the report. Any responses also should be posted in the SI website database. When peer reviews are conducted by FACs, the peer review report and the Agency's response to the committee also are posted on the advisory committee's website.

For products that are not considered "influential" (those categorized as "other"), the Agency may disclose the peer review report and Agency's response to the report (if prepared). Information on the peer review of products not considered influential are not posted on EPA's Peer Review Agenda website.

6.3.3. How Might Peer Review Comments Impact the Work Product?

Peer review comments, when appropriate, enhance the quality of the information EPA disseminates by ensuring that the information that the Agency uses to support and carry out its mission is reliable, accurate and unbiased (i.e., is objective) and that it is appropriate for its intended use (i.e., has utility). A variety of changes to a work product may result from the comments provided during peer review:

- Peer review comments and recommendations may entail significant impacts on the planned project schedule, budget or other resource needs. Management decisions to adjust one or more of these areas may be appropriate.

- The substantive issues or concerns expressed by peer reviewers may suggest that wider scientific and technical consultation is needed to ensure the adequacy of the work product relative to its intended use. If the Agency agrees with the reviewers, additional resources and an extended delivery schedule may be necessary.

- Peer review comments may lead to a better or more thorough analysis, a different interpretation of the results or a different perspective on a topic.

- The peer review comments and recommendations on a draft final product may provide a basis for bringing the associated project to closure.

6.3.4. What Should the Final Work Product Say About the Peer Review Process?

A final peer-reviewed work product *may* include a brief description of the peer review process (e.g., a statement regarding public participation and names/affiliations of the peer reviewers). Frequently, this will be part of a description of the process of developing the product, which can be included in an introduction, preamble or appendix. For ISI and HISAs that support rulemaking, the peer review should be discussed in the preamble of the rule.

When there are significant peer review comments, particularly if they are not accepted, a discussion of the issues and reasons for the Agency's choices should be included in the work product. The level of detail provided is a matter of judgment and should reflect the significance and degree of controversy surrounding the issue.

If ISI or a HISA has not been peer reviewed, this fact should be noted in the document, perhaps in an introduction or description of its scope. This section should briefly indicate the reasons that peer review was not conducted.

6.4. Finalizing the Work Product: When Is the Peer Review of a Work Product Complete?

Performance of the formal peer review is not the final stage in the product's development. Rather, it is an important stage in its development, with the final version (addressing comments) representing the true end of the peer review. The peer review process closes with three major activities:

1. Evaluating peer review comments and recommendations.

2. Utilizing peer review comments for completing the final document or conducting another review, if appropriate.

3. Completing the peer review record (for ISI and HISAs, this includes completing the entry in the SI).

Careful attention to all of these elements, singly and together, ensures a credible and transparent peer review process. Conversely, inattention to detail can nullify the peer review effort. A well-planned peer review applied to a quality draft work product and followed by responsible employment of peer review suggestions in the final product ensures a credible and defensible product for use in Agency decision making. Sometimes the draft work product may not be finalized after the peer review. In these cases, the Agency may decide not to disseminate the Peer Review Report and/or the EPA Response to the Peer Review Report (if any).

Note: For the purposes of the EPA Annual Peer Review Report to OMB, peer review of an influential work product (ISI or HISA) is considered complete when the Agency receives the peer reviewers' final comments (e.g., the peer review report) and the comments are publicly available through the SI.

6.5. The Peer Review Record

6.5.1. What Is the Peer Review Record?

The peer review record is the PRL's formal record (file) of decision on the conduct of the peer review (either internal or external). It includes sufficient documentation (electronic and/or paper) for an uninvolved individual to understand the review process and the outcome. It is the responsibility of the PRL to create a separate review record that may be kept within the overall file for the development of the work product. Once the peer review is completed, it is the responsibility of the PRL to ensure that the peer review record is maintained in accordance with the organization's document retention procedures.

If ISI or a HISA has not been peer reviewed, a record should be created explaining why the product was not peer reviewed, including documentation signed by the DM during the peer review planning process (see Exhibit 2). Some Agency documents, such as strategic plans or analytic blueprints, are not subject to the EPA's Peer Review Policy and do not require peer review; in these cases, no record explaining why the product was not peer reviewed is necessary (see Section 3.3 to determine which work products do not require peer review).

For ISI and HISAs, some of the information from the PRL's official peer review record (e.g., the charge and the draft work product) is entered into the SI database that serves as the primary public interface for these records (see Sections 7.3.2 and 7.3.4). The resulting SI database entry is publicly accessible on the Agency's Peer Review Agenda website[34] through a link to the SI. Since the record in the SI does not contain all the information regarding the peer review, it is not the official peer review record.

6.5.2. What Should Be in the Peer Review Record?

Contents of the peer review record may vary, depending on the type of review undertaken. Documentation should be commensurate with the type of work product and its intended use. Such materials typically include:

- An approved plan specifying the type of peer review;

- Peer review documentation/checklist(s) that contain the rationale for the work product categorization and the signature of the DM approving the categorization (see Exhibit 1);

---

[34] EPA. 2015. *Peer Review Agenda.* http://cfpub.epa.gov/si/si_public_pr_agenda.cfm.

- The draft work product submitted for peer review;

- The materials and information (including the charge) given to the peer reviewers;

- Information about the peer reviewers (e.g., names, affiliations, signed COI forms for each reviewer or a statement concerning potential COIs and their resolution, relevant correspondence);

- Logistical information about the conduct of the peer review (such as times and locations of meetings, if applicable);

- The peer review report, which include reviewers' comments and responses to charge questions;

- A memorandum or other written record, approved by the DM or DM designee, responding to the peer review comments and specifying either acceptance or rebuttal and non-acceptance (when prepared);

- The final work product, including any revisions resulting from the peer review;

- Documentation of any opportunities for public comment, including docket information, if applicable; and

- For ISI and HISAs, SI reference information (e.g., record number).

When deciding if particular materials should be included in the record, the PRL should consider whether the materials would help reconstruct the peer review process and outcome at a later time. If the materials might be helpful, they should be part of the peer review record.

The peer review record is considered complete when it contains a copy of the final work product (when there is one) that addresses the peer review comments, as well as a copy of the Agency's response to the comments (when there is one), including any that were not incorporated.

6.5.3. When Should the Peer Review Record-Building Process Begin?

An early start to developing and maintaining a peer review record will help ensure that the record is complete and helpful. Preferably, the record should begin at the start of the planning stage, once the decision to peer review the work product is made and the product categorization (ISI, HISA or other) is determined and documented.

6.5.4. What Types of Documentation Should Be Maintained When Categorizing Work Products and Determining the Peer Review Mechanism?

When making the determination if a work product is influential and what type of peer review mechanism should be used, these decisions should be documented and include the following: work product peer review categorization, the rationale for the categorization, the peer review mechanism selected and approval by the DM. The flowcharts and checklists found in the Roadmap at the front of this Handbook are tools for assisting the PRLs in evaluating what decisions are needed and how they should be documented. Other tools and products to enhance the transparency and reporting of peer reviews are summarized in Table 1. Individual EPA offices maintain decision documentation for their scientific and technical work products categorized as influential.

6.5.5. How Can the Peer Review Record Improve the Peer Review Process?

A good peer review record supports the planning process and ensures that appropriate peer review is conducted. Also, it permits a retrospective examination of the peer review, and it helps the Agency make appropriate use of peer review comments. In addition, a good record helps ensure that the EPA's Peer Review Policy is implemented. The PRL is responsible for ensuring that the documentation for the peer review record for individual work products is collected and maintained.

6.5.6. What Happens to a Peer Review Record That Pertains to a Rulemaking Action?

The PRL should coordinate with the Federal Docket Management System to ensure that proper docketing procedures are followed for a peer review of a work product supporting a rule. If EPA relies on ISI or a HISA to support a regulatory action, the preamble should include a discussion of how EPA implemented the provisions of the OMB Peer Review Bulletin. See Appendix D, Sound Science and Peer Review in Rulemaking Policy, for a template to use for this purpose.

6.5.7. Are there Differences in Record-Keeping between a Review by Individuals and One by a Panel?

Generally, the content of the two peer review records would be similar. In the case of a review by individuals, such as a letter review, the peer review record typically would contain each individual's comments. For a panel review, the record typically contains a summary or other synthesis of the panel's peer review comments and recommendations (i.e., their peer review report).

6.5.8. Are Internal Peer Review Comments Included in the Peer Review Record?

Comments from formally conducted internal EPA peer reviews should be documented and included in the peer review record. This process does not substitute for Agency clearance. Informal input from EPA colleagues and input from Agency personnel helping to develop the work product need not be included.

Note: An internal EPA peer review may be followed by a separate external peer review. In such a case, the external peer review will stand as the official peer review record, because it is viewed as more independent in nature, may have broader fields of available expertise which can be brought to bear on the issues, and often includes greater depth for specific disciplines.

6.5.9. Where Should the Peer Review Record Be Kept and for How Long?

During the active conduct of the peer review, the PRL is responsible for maintaining the onsite record until the peer review is complete. Once completed, the peer review record should be maintained onsite by the PRL until at least 1 year after the completed peer review is reported in the next annual reporting cycle. The location of the record should be readily identifiable so that interested parties can locate and obtain materials easily and quickly. The peer review record may be kept with other records relating to the overall project as long as it is easily and separately identifiable.

Establishment and maintenance of the archive where the peer review records ultimately reside are an organization's responsibility (i.e., not that of an individual PM or PRL). The PRL should collect the applicable materials and submit them for archiving in accordance with the applicable records-retention schedule(s).

PRLs should consult with their EPA Records Liaison Officer or the EPA's National Records Management Program (http://www.epa.gov/records) to determine the appropriate retention schedule for a peer review record, whether in electronic or paper form. A peer review record may be covered by one or more of the EPA's records-retention schedules. Some peer review records are permanent (e.g., records created by FACs, Integrated Risk Information System [IRIS] peer reviews and Health and Environment Assessment Program Files).

The peer review of products that meet OMB's definitions for ISI or HISAs must be reported and tracked in the EPA's SI (http://cfpub.epa.gov/si/). The SI database entry for such work products should be completed and updated in accordance with the appropriate Agency procedures. When peer review is provided by a FAC, such as the SAB, committee records are created and maintained by the EPA DFO and made available to the public on advisory committee Web pages.[35, 36]

Public dockets serve as the repository for peer review information related to rulemaking (regulatory dockets) or other non-rulemaking actions (general dockets). The appropriate peer review information, however, also should be entered in the SI. There are specific procedures regarding the establishment and use of public dockets for retaining records associated with federal rulemaking and other Agency actions. If a peer review record is included in an EPA docket to support a rulemaking or other Agency action, the Federal Records Act record-retention schedule for dockets must be followed. For details on the EPA's record-retention schedule for dockets, see http://www.epa.gov/records/policy/schedule/.

---

[35] EPA. 2015. *Science Advisory Board.* www.epa.gov/sab.

[36] EPA. 2015. *EPA Clean Air Scientific Advisory Committee (CASAC).* www.epa.gov/casac.

7. Transparency in Peer Review: Public Participation and Reporting

7.1. Overview

The EPA is committed to the independent review of the Agency's scientific products and consistent implementation of its Peer Review Policy across the Agency. Transparency and openness are key objectives of its peer review process (Figure 8). To ensure transparency, the Agency often provides opportunities for participation by the general public, stakeholders and the larger scientific community in the peer review of influential scientific information (ISI). In addition, EPA makes peer review materials (e.g., the peer review plan, the peer review report) for Highly Influential Scientific Assessments (HISAs) and ISI publicly available at the EPA Peer Review Agenda[33] website. Through *Federal Register* notices, website postings and other means, EPA keeps the public informed of its peer review activities. The EPA also submits annual reports on the peer review of influential work products to the Office of Management and Budget (OMB).

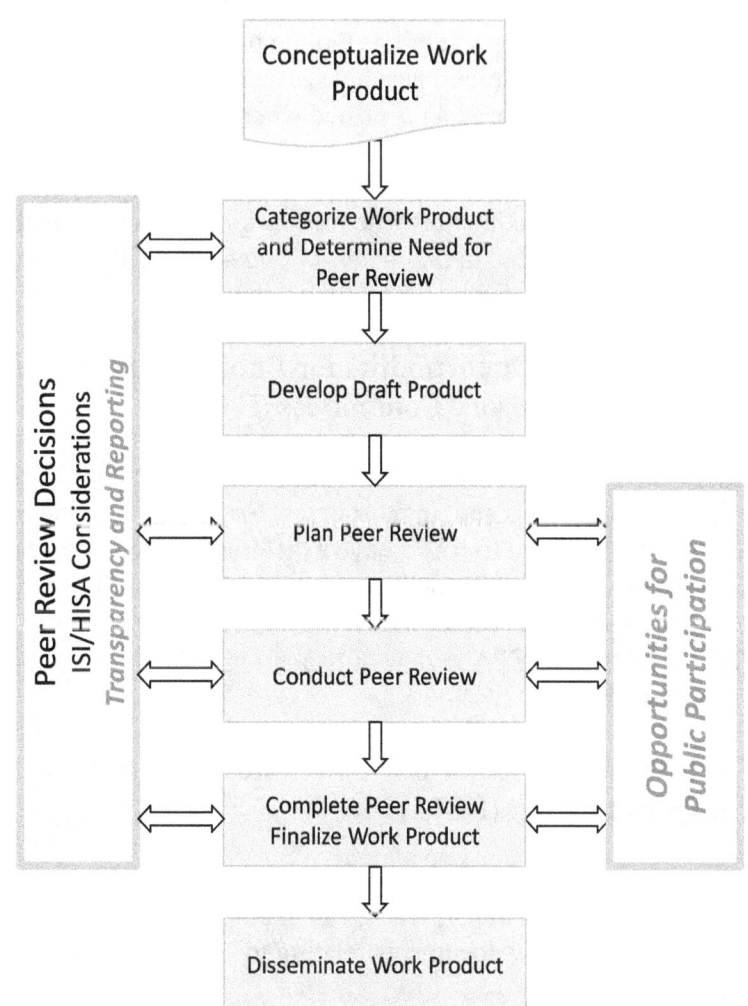

Figure 8. The Peer Review Process: Public Participation and Reporting

7.2. Opportunities for Public Participation

7.2.1. What Are the Opportunities for Public Participation in Peer Review?

The Agency provides opportunities for public participation in its peer review whenever feasible and appropriate. Opportunities are communicated by several means, including the EPA Peer Review

Agenda,[37] relevant Agency websites and *Federal Register* notices. Public comments may be submitted to the Agency in writing, as oral statements during public meetings when panels are convened, or both.

Peer review plans for work products categorized as ISI and HISAs are publicly available on the EPA Peer Review Agenda, and the public may comment on the adequacy of those plans. The Agency indicates in each plan whether the public will have the opportunity to comment on the work product (and if so, how and when opportunities will be provided) and whether the public will be asked to nominate peer reviewers. Sometimes the charge to the peer reviewers is posted for public comment, and for HISAs, the draft work product is posted whenever feasible and appropriate.

For peer reviews conducted by panels selected and managed by an independent contractor, the public may nominate experts and later provide feedback on potential panel members (see Section 4.6.4). These opportunities are announced in *the Federal Register* and EPA may utilize a public docket (at www.regulations.gov) for submission of the comments.

7.2.2. What Are the Opportunities for Public Participation for Peer Reviews Conducted by Federal Advisory Committees (FACs)?

The Federal Advisory Committee Act (FACA) requires that the public have an opportunity to provide written comments to FACs and, in most cases, FACs also provide opportunities for oral comments.[38] Public comments provided to FACs have a different purpose than public comment provided to EPA offices because they inform the deliberations of the FAC as it reviews the draft EPA work product. Members of the public can submit relevant comments pertaining to the group providing advice, the EPA's charge questions, EPA review or background documents, and draft advisory reports prepared by a FAC or its panels.

7.2.3. Is Information Regarding a Peer Review Subject to Release Under the Freedom of Information Act (FOIA)?

Information regarding a peer review is subject to release if EPA receives a FOIA request unless the peer review information meets the criteria for an exemption under the FOIA 5 U.S.C. § 552(b). It should be noted, however, that many documents relating to Agency peer reviews are available to the public on the EPA website.

7.3. Reporting on Peer Reviews

7.3.1. What Are the EPA's Reporting Practices?

As part of the EPA's systematic planning process, EPA publicly posts information on the peer review activities of EPA's forthcoming influential (HISA and ISI) scientific product disseminations on the EPA's Peer Review Agenda (PRA) website.[39] Pursuant to the OMB Peer Review Bulletin, for each entry on the PRA, the Agency provides a link to the peer review plan, the charge to the reviewers, the peer review report, the names and affiliations of the peer reviewers (in the peer review report or as a

---

[37] EPA. 2015. *Peer Review Agenda.* http://cfpub.epa.gov/si/si_public_pr_agenda.cfm.

[38] 5 U.S.C. App. 2 Section 10(a)(3) and § 102–3.140(c) and (d) of the U.S. General Services Administration FAC Management final rule.

[39] EPA. 2015. *Peer Review Agenda.* http://cfpub.epa.gov/si/si_public_pr_agenda.cfm.

separate file), any Agency response to comments, and, whenever feasible and appropriate, the draft work product for HISAs.

The PRA is a component of the EPA Science Inventory (SI), a searchable database of scientific and technical work products developed by EPA and accessible at www.epa.gov/si. Product metadata including peer review information and related documents, are entered into the SI and then published to the Agency PRA, which is also linked to the U.S. Government's official web portal FirstGov at http://www.FirstGov.gov.

In addition to reporting on peer review through the PRA, the Agency officially submits an annual report on peer review to OMB that summarizes all the external peer reviews of HISA and ISI products that were completed during the fiscal year. For the purposes of reporting to OMB, "completed" is defined as having received the peer review report (see Sections 6.4.1 and 7.4). The Agency response to the peer reviewer comments is also reported to OMB in the annual report, among other information.

EPA offices also communicate their peer review activities through press releases, website postings, dockets and *Federal Register* notices (see Appendix I for example notices).

7.3.2. What Information Should Be Provided in the Science Inventory Peer Review Plan Regarding ISI and HISAs?

Since EPA allows the public to view and comment on the Agency's peer review plans for activities or products categorized as ISI or HISAs, the following information should be provided for each activity or product into the SI:

- A paragraph including the title, subject and purpose of the activity or product.

- An Agency contact to whom inquiries may be directed to learn the specifics of the peer review plan.

- The categorization of the work product (e.g., ISI, HISA).

- The timing of the review (including any deferrals).

- The process by which the review will be conducted (e.g., a panel or individual letter review, an alternative procedure).

- Opportunities for the public to comment on the work product to be peer reviewed, including how and when these opportunities will be provided, if applicable.

- Any significant and relevant public comments that EPA will provide to the peer reviewers before they conduct their review.

- A succinct description of the primary disciplines or expertise needed in the peer review.

- The number of peer reviewers expected.

- The organization that will select the reviewers (e.g., EPA, a designated outside organization).

- Opportunities for the public, including scientific or professional societies, to nominate peer reviewers, if applicable.

After new or updated product and peer review information is entered into the SI, the SI Product Coordinator or his/her designee generates (from within the SI) a public peer review plan and posts the plan to the Peer Review Agenda website. The SI Product Coordinator should also post or link other relevant peer review documents to the PRA from the SI.

EPA offices are expected to keep this information current by updating agenda entries of influential work products at least every 6 months. Real-time updates occasionally may be necessary, for example, when there is an imminent change in the timing for the peer review of a high-visibility work product or a change in the timing of the public availability of a draft of a HISA.

7.3.3. Which Products Generated Under EPA Grants or Cooperative Agreements Should Be Reported in the Science Inventory?

As a matter of practice, EPA organizations are encouraged, but not required, to include in the SI those scientific and technical work products that are produced under grants and cooperative agreements so that EPA staff and the public are aware of the ongoing work. If a grant or cooperative agreement product is likely to be used in Agency decision making (assuming this use is incidental to the principal purpose of the agreement), it generally should be considered a candidate for peer review and noted as such in the SI by the Peer Review Coordinator (PRC).

7.3.4. Does the Agency Report on Peer Review of Scientific and Technical Work Products That Are Not ISI or HISAs?

Each EPA office is responsible for reporting peer-reviewed work products categorized as "other" upon request. For example, a list of these work products may be requested from each EPA office for inclusion in annual reports (e.g., for the Agency's Annual Report on Scientific Integrity) and for the purposes of monitoring compliance with the EPA's Peer Review Policy and this Handbook. Offices are encouraged to include information in the SI on the peer review of these other work products not categorized as ISI or HISAs but are not required to do so.

7.4. Annual Report to OMB on EPA Peer Reviews

The EPA submits an annual report to OMB that summarizes the peer reviews of all of the ISI and HISAs that were conducted during the previous fiscal year. Release of any reviewer information retrieved by a personal identifier will be performed in accordance with the Privacy Act of 1974 (5 U.S.C. § 552a, as amended), as interpreted in OMB implementing guidance, 40 *Fed. Reg.* 28,948 (Jul. 9, 1975). The OMB Peer Review Bulletin states that the annual report should include the following:

- The number of peer reviews conducted subject to the OMB Peer Review Bulletin.

- The number of times alternative procedures were invoked (see Section IV of the OMB Peer Review Bulletin).

- The number of times waivers or deferrals were invoked and, in the case of deferrals, the length of time elapsed between the deferral and the peer review.

- Any decision to appoint a peer reviewer pursuant to any exception to the applicable independence or conflict of interest (COI) standards of the OMB Peer Review Bulletin.

- The number of peer review panels that were conducted in public and the number that included public comment.

- The number of public comments provided on the peer review plans.

- The number of peer reviewers used who were recommended by professional societies.

APPENDIX A. EPA PEER REVIEW POLICY

UNITED STATES ENVIRONMENTAL PROTECTION AGENCY
WASHINGTON, D.C. 20460

JAN 31 2006

THE ADMINISTRATOR

MEMORANDUM

SUBJECT: Peer Review Program

TO: Assistant Administrators
     General Counsel
     Inspector General
     Associate Administrators
     Regional Administrators
     Staff Office Directors

We have made tremendous strides in improving our peer review program at EPA since the Agency's Peer Review Policy was reaffirmed in 1994. Today I am updating the Peer Review Policy to emphasize the critical role that peer review plays in our efforts to ensure that EPA's decisions rest on sound, credible science and data (see attached policy statement).

Peer review at EPA takes several different forms, ranging from informal consultations with Agency colleagues who were not involved in developing the product to the formal, public processes of the Science Advisory Board (SAB) and the FIFRA Scientific Advisory Panel (SAP). In any form, peer review assists EPA's work by bringing independent expert experience and judgment to bear on issues before the Agency to the benefit of the final product.

In 1994 the Science Policy Council (SPC) and its Steering Committee were asked to undertake an initiative to ensure that EPA has a comprehensive Agency-wide program for implementing its Peer Review Policy. I commend the SPC for its diligence and success in meeting this objective. The SPC has made substantial improvements in the Peer Review Handbook, sponsored training of Agency managers and staff in peer review procedures, identified scientific and technical work products that merit peer review, and developed a publicly available data base of the peer review activities across the Agency. EPA has a strong and well-recognized peer review program as a direct result of these efforts.

In 2004 the Office of Management and Budget (OMB) issued a "Final Information Quality Bulletin for Peer Review" that contains provisions for peer review at all federal agencies. The OMB Bulletin applies to influential scientific information and highly influential scientific assessments. The SPC has updated the Agency's Peer Review Handbook, in part to incorporate the provisions of the OMB Bulletin, and to reflect the experience gained from implementing the program over the last decade.

I ask that you continue to implement fully the provisions of our Peer Review Policy, and I expect the Science Policy Council to continue its role in overseeing and strengthening EPA's peer review program. We must ensure that our decisions are based on the highest quality, peer-reviewed scientific and technical information.

Stephen L. Johnson

Attachment

cc: Science Policy Council
       Science Policy Council Steering Committee

PEER REVIEW AND PEER INVOLVEMENT
AT THE U.S.ENVIRONMENTAL PROTECTION AGENCY

This document establishes the policy of the United States Environmental Protection Agency (EPA) for peer review of scientifically and technically based work products, including economic and social science products, that are intended to inform Agency decisions. Peer review, a form of *peer involvement,* is one process through which EPA staff augment their capabilities by inviting independent subject-matter experts to provide objective evaluation of the work product.

PEER REVIEW

EPA strives to ensure that the scientific and technical bases of its decisions meet two important criteria: (1) they are based upon the best current knowledge from science, engineering, and other domains of technical expertise; and (2) they are credible. Peer review, a process based on the principles of obtaining the best technical and scientific expertise with appropriate independence, is central to sound science and helps the Agency meet these important criteria. Peer review occurs when scientifically and technically based work products are evaluated by relevant experts who were not involved in creating the product. Properly applied, peer review not only enriches the quality of work products but also adds a degree of credibility that cannot be achieved in any other way. Furthermore, peer review early in the development of work products in some cases may conserve future resources by steering the development along the most efficacious course.

Peer review generally takes one of two approaches:

- Internal, in which the reviewers are independent experts from inside EPA.

- External, in which the reviewers are independent experts from outside EPA.

POLICY STATEMENT

Peer review of all scientific and technical information that is intended to inform or support Agency decisions is encouraged and expected. Influential scientific information, including highly influential scientific assessments, should be peer reviewed in accordance with the Agency's Peer Review Handbook. All Agency managers are accountable for ensuring that Agency policy and guidance are appropriately applied in determining if their work products are influential or highly influential, and for deciding the nature, scope, and timing of their peer review. For highly influential scientific assessments, external peer review is the expected procedure. For influential scientific information intended to support important decisions, or for work products that have special importance in their own right, external peer review is the approach of choice. Peer review is not restricted to the nearly final version of work products; in fact, peer review at the planning stage can often be extremely beneficial.

LEGAL EFFECT

This policy statement does not establish or affect legal rights or obligations. Rather, it confirms the importance of peer review where appropriate, outlines relevant principles, and identifies factors Agency staff should consider in implementing the policy. On a continuing basis, Agency management is expected to evaluate the policy as well as the results of its application throughout the Agency and undertake revisions as necessary. Therefore, the policy does not stand alone; nor does it establish a binding norm that is finally determinative of the issues addressed.

IMPLEMENTATION

The Science Policy Council is responsible for overseeing Agency-wide implementation of this policy, including: promoting consistent interpretation; assessing Agency-wide progress; developing recommendations for revisions of the policy as necessary; and issuing the *Peer Review Handbook,* which provides additional information and procedures on implementing this policy. Assistant Administrators, Regional Administrators, and other senior managers remain ultimately responsible for ensuring the appropriate application of Agency policy and guidance in identifying work products subject to peer review, determining the type and timing of such review, documenting the process and outcome of each peer review, ensuring that the Science Inventory is kept current, and otherwise implementing the policy within their organizational units.

The policy is effective immediately.

APPROVED: _____ DATE: JAN 3 1 2006
STEPHEN L. JOHNSON, ADMINISTRATOR

APPENDIX B. OMB INFORMATION QUALITY BULLETIN
FOR PEER REVIEW

This appendix contains the text of the OMB Information Quality Bulletin for Peer Review.

December 15, 2004

OFFICE OF MANAGEMENT AND BUDGET
Final Information Quality Bulletin for Peer Review

INTRODUCTION

This Bulletin establishes that important scientific information shall be peer reviewed by qualified specialists before it is disseminated by the federal government. We published a proposed Bulletin on September 15, 2003. Based on public comments, we published a revised proposal for additional comment on April 28, 2004. We are now finalizing the April version, with minor revisions responsive to the public's comments.

The purpose of the Bulletin is to enhance the quality and credibility of the government's scientific information. We recognize that different types of peer review are appropriate for different types of information. Under this Bulletin, agencies are granted broad discretion to weigh the benefits and costs of using a particular peer review mechanism for a specific information product. The selection of an appropriate peer review mechanism for scientific information is left to the agency's discretion. Various types of information are exempted from the requirements of this Bulletin, including time-sensitive health and safety determinations, in order to ensure that peer review does not unduly delay the release of urgent findings.

This Bulletin also applies stricter minimum requirements for the peer review of highly influential scientific assessments, which are a subset of influential scientific information. A scientific assessment is an evaluation of a body of scientific or technical knowledge that typically synthesizes multiple factual inputs, data, models, assumptions, and/or applies best professional judgment to bridge uncertainties in the available information. To ensure that the Bulletin is not too costly or rigid, these requirements for more intensive peer review apply only to the more important scientific assessments disseminated by the federal government.

Even for these highly influential scientific assessments, the Bulletin leaves significant discretion to the agency formulating the peer review plan. In general, an agency conducting a peer review of a highly influential scientific assessment must ensure that the peer review process is transparent by making available to the public the written charge to the peer reviewers, the peer reviewers' names, the peer reviewers' report(s), and the agency's response to the peer reviewers' report(s). The agency selecting peer reviewers must ensure that the reviewers possess the necessary expertise. In addition, the agency must address reviewers' potential conflicts of interest (including those stemming from ties to regulated businesses and other stakeholders) and independence from the agency. This Bulletin requires agencies to adopt or adapt the committee selection policies employed by the National Academy of Sciences (NAS)[1] when selecting peer reviewers who are not government employees. Those that are government employees are subject to federal ethics requirements. The use of a transparent process, coupled with the selection of qualified and independent peer reviewers, should improve the quality of government science while promoting public confidence in the integrity of the government's scientific products.

PEER REVIEW

Peer review is one of the important procedures used to ensure that the quality of published information meets the standards of the scientific and technical community. It is a form of deliberation involving an exchange of judgments about the appropriateness of methods and the strength of the author's inferences.[2] Peer review involves the review of a draft product for quality by specialists in the field who were not involved in producing the draft.

The peer reviewer's report is an evaluation or critique that is used by the authors of the draft to improve the product. Peer review typically evaluates the clarity of hypotheses, the validity of the research design, the quality of data collection procedures, the robustness of the methods employed, the appropriateness of the methods for the

---

[1] National Academy of Sciences, "Policy and Procedures on Committee Composition and Balance and Conflicts of Interest for Committees Used in the Development of Reports.," May 2003; Available at: http://www.nationalacademies.org/coi/index.html

[2] Carnegie Commission on Science, Technology, and Government, Risk and the Environment: Improving Regulatory Decision Making, Carnegie Commission, New York, 1993: 75.

hypotheses being tested, the extent to which the conclusions follow from the analysis, and the strengths and limitations of the overall product.

Peer review has diverse purposes. Editors of scientific journals use reviewer comments to help determine whether a draft scientific article is of sufficient quality, importance, and interest to a field of study to justify publication. Research funding organizations often use peer review to evaluate research proposals. In addition, some federal agencies make use of peer review to obtain evaluations of draft information that contains important scientific determinations.

Peer review should not be confused with public comment and other stakeholder processes. The selection of participants in a peer review is based on expertise, with due consideration of independence and conflict of interest. Furthermore, notice-and- comment procedures for agency rulemaking do not provide an adequate substitute for peer review, as some experts -- especially those most knowledgeable in a field -- may not file public comments with federal agencies.

The critique provided by a peer review often suggests ways to clarify assumptions, findings, and conclusions. For instance, peer reviews can filter out biases and identify oversights, omissions, and inconsistencies.[3] Peer review also may encourage authors to more fully acknowledge limitations and uncertainties. In some cases, reviewers might recommend major changes to the draft, such as refinement of hypotheses, reconsideration of research design, modifications of data collection or analysis methods, or alternative conclusions. However, peer review does not always lead to specific modifications in the draft product. In some cases, a draft is in excellent shape prior to being submitted for review. In others, the authors do not concur with changes suggested by one or more reviewers.

---

[3] William W. Lowrance, <u>Modern Science and Human Values</u>., Oxford University Press, New York, NY 1985: 85.

Peer review may take a variety of forms, depending upon the nature and importance of the product. For example, the reviewers may represent one scientific discipline or a variety of disciplines; the number of reviewers may range from a few to more than a dozen; the names of each reviewer may be disclosed publicly or may remain anonymous (e.g., to encourage candor); the reviewers may be blinded to the authors of the report or the names of the authors may be disclosed to the reviewers; the reviewers may prepare individual reports or a panel of reviewers may be constituted to produce a collaborative report; panels may do their work electronically or they may meet together in person to discuss and prepare their evaluations; and reviewers may be compensated for their work or they may donate their time as a contribution to science or public service.

For large, complex reports, different reviewers may be assigned to different chapters or topics. Such reports may be reviewed in stages, sometimes with confidential reviews that precede a public process of panel review. As part of government-sponsored peer review, there may be opportunity for written and/or oral public comments on the draft product.

The results of peer review are often only one of the criteria used to make decisions about journal publication, grant funding, and information dissemination. For instance, the editors of scientific journals (rather than the peer reviewers) make final decisions about a manuscript's appropriateness for publication based on a variety of considerations. In research-funding decisions, the reports of peer reviewers often play an important role, but the final decisions about funding are often made by accountable officials based on a variety of considerations. Similarly, when a government agency sponsors peer review of its own draft documents, the peer review reports are an important factor in information dissemination decisions but rarely are the sole consideration. Agencies are not expected to cede their discretion with regard to dissemination or use of information to peer reviewers; accountable agency officials must make the final decisions.

There are a multiplicity of science advisory procedures used at federal agencies and across the wide variety of scientific products prepared by agencies.[4] In response to congressional inquiry, the U.S. General Accounting Office (now the Government Accountability Office) documented the variability in both the definition and implementation of peer review across agencies.[5] The Carnegie Commission on Science, Technology and Government[6] has highlighted the importance of "internal" scientific advice (within the agency) and "external" advice (through scientific advisory boards and other mechanisms).

A wide variety of authorities have argued that peer review practices at federal agencies need to be strengthened.[7] Some arguments focus on specific types of scientific products (e.g., assessments of health, safety and environmental hazards).[8] The Congressional/Presidential Commission on Risk Assessment and Risk Management suggests that "peer review of economic and social science information should have as high a priority as peer review of health, ecological, and engineering information."[9]

---

[4] Sheila Jasanoff, The Fifth Branch: Science Advisors as Policy Makers, Harvard University Press, Boston, 1990.

[5] U.S. General Accounting Office, Federal Research: Peer Review Practices at Federal Agencies Vary, GAO/RCED-99-99, Washington, D.C., 1999.

[6] Carnegie Commission on Science, Technology, and Government, Risk and the Environment: Improving Regulatory Decision Making, Carnegie Commission, New York, 1993: 90.

[7] National Academy of Sciences, Peer Review in the Department of Energy – Office of Science and Technology, Interim Report, National Academy Press, Washington, D.C., 1997; National Academy of Sciences, Peer Review in Environmental Technology Development: The Department of Energy – Office of Science and Technology, National Academy Press, Washington, D.C., 1998; National Academy of Sciences, Strengthening Science at the U.S. Environmental Protection Agency: Research-Management and Peer-Review Practices, National Academy Press, Washington, D.C. 2000; U.S. General Accounting Office, EPA's Science Advisory Board Panels: Improved Policies and Procedures Needed to Ensure Independence and Balance, GAO-01-536, Washington, D.C., 2001; U. S. Environmental Protection Agency, Office of Inspector General, Pilot Study: Science in Support of Rulemaking 2003-P-00003, Washington, D.C., 2002; Carnegie Commission on Science, Technology, and Government, In the National Interest: The Federal Government in the Reform of K-12 Math and Science Education, Carnegie Commission, New York, 1991; U.S. General Accounting Office, Endangered Species Program: Information on How Funds Are Allocated and What Activities are Emphasized, GAO-02-581, Washington, D.C. 2002.

[8] National Research Council, Science and Judgment in Risk Assessment, National Academy Press, Washington, D.C., 1994.

[9] Presidential/Congressional Commission on Risk Assessment and Risk Management, Risk Commission Report, Volume 2, Risk Assessment and Risk Management in Regulatory Decision-Making, 1997:103.

Some agencies have formal peer review policies, while others do not. Even agencies that have such policies do not always follow them prior to the release of important scientific products.

Prior to the development of this Bulletin, there were no government-wide standards concerning when peer review is required and, if required, what type of peer review processes are appropriate. No formal interagency mechanism existed to foster cross- agency sharing of experiences with peer review practices and policies. Despite the importance of peer review for the credibility of agency scientific products, the public lacked a consistent way to determine when an important scientific information product is being developed by an agency, the type of peer review planned for that product, or whether there would be an opportunity to provide comments and data to the reviewers.

This Bulletin establishes minimum standards for when peer review is required for scientific information and the types of peer review that should be considered by agencies in different circumstances. It also establishes a transparent process for public disclosure of peer review planning, including a web-accessible description of the peer review plan that the agency has developed for each of its forthcoming influential scientific disseminations.

LEGAL AUTHORITY FOR THE BULLETIN

This Bulletin is issued under the Information Quality Act and OMB's general authorities to oversee the quality of agency information, analyses, and regulatory actions. In the Information Quality Act, Congress directed OMB to issue guidelines to "provide policy and procedural guidance to Federal agencies for ensuring and maximizing the quality, objectivity, utility and integrity of information" disseminated by Federal agencies. Pub. L. No. 106-554, § 515(a). The Information Quality Act was developed as a supplement to the Paperwork Reduction Act, 44 U.S.C. § 3501 et seq., which requires OMB, among other things, to "develop and oversee the implementation of policies, principles, standards, and guidelines to apply to Federal agency dissemination of public information." In addition, Executive Order 12866, 58 Fed. Reg. 51,735 (Oct. 4, 1993), establishes that OIRA is "the repository of expertise concerning regulatory issues," and it directs OMB to provide guidance to the agencies on regulatory planning. E.O. 12866, § 2(b). The Order also requires that "[e]ach agency shall base its decisions on the best reasonably obtainable scientific, technical, economic, or other information." E.O. 12866, § 1(b)(7). Finally, OMB has authority in certain circumstances to manage the agencies under the purview of the President's Constitutional authority to supervise the unitary Executive Branch. All of these authorities support this Bulletin.

THE REQUIREMENTS OF THIS BULLETIN

This Bulletin addresses peer review of scientific information disseminations that contain findings or conclusions that represent the official position of one or more agencies of the federal government.

Section I: Definitions

Section I provides definitions that are central to this Bulletin. Several terms are identical to or based on those used in OMB's government-wide information quality guidelines, 67 Fed. Reg. 8452 (Feb. 22, 2002), and the Paperwork Reduction Act, 44 U.S.C. § 3501 et seq.

The term "Administrator" means the Administrator of the Office of Information and Regulatory Affairs in the Office of Management and Budget (OIRA).

The term "agency" has the same meaning as in the Paperwork Reduction Act, 44 U.S.C. § 3502(1).

The term "Information Quality Act" means Section 515 of Public Law 106-554 Pub. L. No. 106-554, § 515, 114 Stat. 2763, 2763A-153-154 (2000)).

The term "dissemination" means agency initiated or sponsored distribution of information to the public. Dissemination does not include distribution limited to government employees or agency contractors or grantees; intra- or inter-agency use or sharing of government information; or responses to requests for agency records under the Freedom of Information Act, the Privacy Act, the Federal Advisory Committee Act, the Government Performance and Results Act, or similar laws. This definition also excludes distribution limited to correspondence with individuals or persons, press releases, archival records, public filings, subpoenas and adjudicative processes. In the context of this Bulletin, the definition of "dissemination" modifies the definition in OMB's government-wide information quality guidelines to address the need for peer review prior to official dissemination of the information product. Accordingly, under this Bulletin, "dissemination" also excludes information distributed for peer review in compliance with this Bulletin or shared confidentially with scientific colleagues, provided that the distributing agency includes an appropriate and clear disclaimer on the information, as explained more fully below. Finally, the Bulletin does not directly cover information supplied to the government by third parties (e.g., studies by private consultants, companies and private, non-profit organizations, or research institutions such as universities). However, if an agency plans to disseminate information supplied by a third party (e.g., using this information as the basis for an agency's factual determination that a particular behavior causes a disease), the requirements of the Bulletin apply, if the dissemination is "influential".

In cases where a draft report or other information is released by an agency solely for purposes of peer review, a question may arise as to whether the draft report constitutes an official "dissemination" under information-quality guidelines. Section I instructs agencies to make this clear by presenting the following disclaimer in the report:

"THIS INFORMATION IS DISTRIBUTED SOLELY FOR THE PURPOSE OF PRE- DISSEMINATION PEER REVIEW UNDER APPLICABLE INFORMATION QUALITY GUIDELINES. IT HAS NOT BEEN FORMALLY DISSEMINATED BY

[THE AGENCY]. IT DOES NOT REPRESENT AND SHOULD NOT BE CONSTRUED TO REPRESENT ANY AGENCY DETERMINATION OR POLICY."

In cases where the information is highly relevant to specific policy or regulatory deliberations, this disclaimer shall appear on each page of a draft report. Agencies also shall discourage state, local, international and private organizations from using information in draft reports that are undergoing peer review. Draft influential scientific information presented at scientific meetings or shared confidentially with colleagues for scientific input prior to peer review shall include the disclaimer: "THE FINDINGS AND CONCLUSIONS IN THIS REPORT (PRESENTATION) HAVE NOT BEEN FORMALLY DISSEMINATED BY [THE AGENCY] AND SHOULD NOT BE CONSTRUED TO REPRESENT ANY AGENCY DETERMINATION OR POLICY."

An information product is not covered by the Bulletin unless it represents an official view of one or more departments or agencies of the federal government. Accordingly, for the purposes of this Bulletin, "dissemination" excludes research produced by government- funded scientists (e.g., those supported extramurally or intramurally by federal agencies or those working in state or local governments with federal support) if that information is not represented as the views of a department or agency (i.e., they are not official government disseminations). For influential scientific information that does not have the imprimatur of the federal government, scientists employed by the federal government are required to include in their information product a clear disclaimer that "the findings and conclusions in this report are those of the author(s) and do not necessarily represent the views of the funding agency." A similar disclaimer is advised for non-government employees who publish government-funded research.

For the purposes of the peer review Bulletin, the term "scientific information" means factual inputs, data, models, analyses, technical information, or scientific assessments related to such disciplines as the behavioral and social sciences, public health and medical sciences, life and earth sciences, engineering, or physical sciences. This includes any communication or representation of knowledge such as facts or data, in any medium or form, including textual, numerical, graphic, cartographic, narrative, or audiovisual forms. This definition includes information that an agency disseminates from a web page, but does not include the provision of hyperlinks on a web page to information that others disseminate. This definition excludes opinions, where the agency's presentation makes clear that an individual's opinion, rather than a statement of fact or of the agency's findings and conclusions, is being offered.

The term "influential scientific information" means scientific information the agency reasonably can determine will have or does have a clear and substantial impact on important public policies or private sector decisions. In the term "influential scientific information," the term "influential" should be interpreted consistently with OMB's government-wide information quality guidelines and the information quality guidelines of the agency. Information dissemination can have a significant economic impact even if it is not part of a rulemaking. For instance, the economic viability of a technology can be influenced by the government's characterization of its attributes. Alternatively, the federal government's assessment of risk can directly or indirectly influence the response actions of state and local agencies or international bodies.

One type of scientific information is a scientific assessment. For the purposes of this Bulletin, the term "scientific assessment" means an evaluation of a body of scientific or technical knowledge, which typically synthesizes multiple factual inputs, data, models, assumptions, and/or applies best professional judgment to bridge uncertainties in the available information. These assessments include, but are not limited to, state-of-science reports; technology assessments; weight-of-evidence analyses; meta-analyses; health, safety, or ecological risk assessments; toxicological characterizations of substances; integrated assessment models; hazard determinations; or exposure assessments. Such assessments often draw upon knowledge from multiple disciplines. Typically, the data and models used in scientific assessments have already been subject to some form of peer review (e.g., refereed journal peer review or peer review under Section II of this Bulletin).

Section II: Peer Review of Influential Scientific Information

Section II requires each agency to subject "influential" scientific information to peer review prior to dissemination. For dissemination of influential scientific information, Section II provides agencies broad discretion in determining what type of peer review is appropriate and what procedures should be employed to select appropriate reviewers. Agencies are directed to chose a peer review mechanism that is adequate, giving due consideration to the novelty and complexity of the science to be reviewed, the relevance of the information to decision making, the extent of prior peer reviews, and the expected benefits and costs of additional review.

The National Academy of Public Administration suggests that the intensity of peer review should be commensurate with the significance of the information being disseminated and the likely implications for policy decisions.[10] Furthermore, agencies need to consider tradeoffs between depth of peer review and timeliness.[11] More rigorous peer review is necessary for information that is based on novel methods or presents complex challenges for interpretation. Furthermore, the need for rigorous peer review is greater when the information contains precedent-setting methods or models, presents conclusions that are likely to change prevailing practices, or is likely to affect policy decisions that have a significant impact.

This tradeoff can be considered in a benefit-cost framework. The costs of peer review include both the direct costs of the peer review activity and those stemming from potential delay in government and private actions that can result from peer review. The benefits of peer review are equally clear: the insights offered by peer reviewers may lead to policy with more benefits and/or fewer costs. In addition to contributing to strong science, peer review, if performed fairly and rigorously, can build consensus among stakeholders and reduce the temptation for courts and legislators to second-guess or

---

[10] National Academy of Public Administration, Setting Priorities, Getting Results: A New Direction for EPA, National Academy Press, Washington, D.C., 1995:23.

[11] Presidential/Congressional Commission on Risk Assessment and Risk Management, Risk Commission Report, 1997.

overturn agency actions.[12] While it will not always be easy for agencies to quantify the benefits and costs of peer review, agencies are encouraged to approach peer review from a benefit-cost perspective.

Regardless of the peer review mechanism chosen, agencies should strive to ensure that their peer review practices are characterized by both scientific integrity and process integrity. "Scientific integrity," in the context of peer review, refers to such issues as "expertise and balance of the panel members; the identification of the scientific issues and clarity of the charge to the panel; the quality, focus and depth of the discussion of the issues by the panel; the rationale and supportability of the panel's findings; and the accuracy and clarity of the panel report." "Process integrity" includes such issues as "transparency and openness, avoidance of real or perceived conflicts of interest, a workable process for public comment and involvement," and adherence to defined procedures.[13]

When deciding what type of peer review mechanism is appropriate for a specific information product, agencies will need to consider at least the following issues: individual versus panel review; timing; scope of the review; selection of reviewers; disclosure and attribution; public participation; disposition of reviewer comments; and adequacy of prior peer review.

Individual versus Panel Review

Letter reviews by several experts generally will be more expeditious than convening a panel of experts. Individual letter reviews are more appropriate when a draft document covers only one discipline or when premature disclosure of a sensitive report to a public panel could cause harm to government or private interests. When time and resources

---

[12] Mark R. Powell, *Science at EPA: Information in the Regulatory Process, Resources for the Future*, Washington, D.C., 1999: 148, 176; Sheila Jasanoff, *The Fifth Branch: Science Advisors as Policy Makers*, Harvard University Press, Boston, 1990: 242.

[13] ILSI Risk Sciences Institute, "Policies and Procedures: Model Peer Review Center of Excellence," 2002: Available at http://rsi.ilsi.org/file/Policies&Procedures.pdf.

warrant, panels are preferable, as they tend to be more deliberative than individual letter reviews and the reviewers can learn from each other. There are also multi-stage processes in which confidential letter reviews are conducted prior to release of a draft document for public notice and comment, followed by a formal panel review. These more rigorous and expensive processes are particularly valuable for highly complex, multidisciplinary, and more important documents, especially those that are novel or precedent-setting.

Timing of Peer Review

As a general rule, it is most useful to consult with peers early in the process of producing information. For example, in the context of risk assessments, it is valuable to have the choice of input data and the specification of the model reviewed by peers before the agency invests time and resources in implementing the model and interpreting the results. "Early" peer review occurs in time to "focus attention on data inadequacies in time for corrections.

When an information product is a critical component of rule-making, it is important to obtain peer review before the agency announces its regulatory options so that any technical corrections can be made before the agency becomes invested in a specific approach or the positions of interest groups have hardened. If review occurs too late, it is unlikely to contribute to the course of a rulemaking. Furthermore, investing in a more rigorous peer review early in the process "may provide net benefit by reducing the prospect of challenges to a regulation that later may trigger time consuming and resource-draining litigation."[14]

---

[14] Fred Anderson, Mary Ann Chirba Martin, E Donald Elliott, Cynthia Farina, Ernest Gellhorn, John D. Graham, C. Boyden Gray, Jeffrey Holmstead, Ronald M. Levin, Lars Noah, Katherine Rhyne, Jonathan Baert Wiener, "Regulatory Improvement Legislation: Risk Assessment, Cost-Benefit Analysis, and Judicial Review," Duke Environmental Law and Policy Forum, Fall 2000, vol. XI (1): 132.

The "charge" contains the instructions to the peer reviewers regarding the objective of the peer review and the specific advice sought. The importance of the information, which shapes the goal of the peer review, influences the charge. For instance, the goal of the review might be to determine the utility of a body of literature for drawing certain conclusions about the feasibility of a technology or the safety of a product. In this context, an agency might ask reviewers to determine the relevance of conclusions drawn in one context for other contexts (e.g., different exposure conditions or patient populations).

The charge to the reviewers should be determined in advance of the selection of the reviewers. In drafting the charge, it is important to remember the strengths and limitations of peer review. Peer review is most powerful when the charge is specific and steers the reviewers to specific technical questions while also directing reviewers to offer a broad evaluation of the overall product.

Uncertainty is inherent in science, and in many cases individual studies do not produce conclusive evidence. Thus, when an agency generates a scientific assessment, it is presenting its scientific judgment about the accumulated evidence rather than scientific fact.[15] Specialists attempt to reach a consensus by weighing the accumulated evidence. Peer reviewers can make an important contribution by distinguishing scientific facts from professional judgments. Furthermore, where appropriate, reviewers should be asked to provide advice on the reasonableness of judgments made from the scientific evidence.

However, the charge should make clear that the reviewers are not to provide advice on the policy (e.g., the amount of uncertainty that is acceptable or the amount of precaution that should be embedded in an analysis). Such considerations are the purview of the government.[16]

---

[15] Mark R. Powell, Science at EPA: Information in the Regulatory Process, Resources for the Future, Washington, D.C., 1999. 139 http://intranet.epa.gov/ocem/faca/guidance/2012_03_epa_faca_handbook.pdf.

[16] . Ibid.

The charge should ask that peer reviewers ensure that scientific uncertainties are clearly identified and characterized. Since not all uncertainties have an equal effect on the conclusions drawn, reviewers should be asked to ensure that the potential implications of the uncertainties for the technical conclusions drawn are clear. In addition, peer reviewers might be asked to consider value-of-information analyses that identify whether more research is likely to decrease key uncertainties.[17] Value-of-information analysis was suggested for this purpose in the report of the Presidential/Congressional Commission on Risk Assessment and Risk Management.[18] A description of additional research that would appreciably influence the conclusions of the assessment can help an agency assess and target subsequent efforts.

Selection of Reviewers

Expertise. The most important factor in selecting reviewers is expertise: ensuring that the selected reviewer has the knowledge, experience, and skills necessary to perform the review. Agencies shall ensure that, in cases where the document being reviewed spans a variety of scientific disciplines or areas of technical expertise, reviewers who represent the necessary spectrum of knowledge are chosen. For instance, expertise in applied mathematics and statistics is essential in the review of models, thereby allowing an audit of calculations and claims of significance and robustness based on the numeric data.[19]

For some reviews, evaluation of biological plausibility is as important as statistical modeling. Agencies shall consider requesting that the public, including scientific and professional societies, nominate potential reviewers.

---

[17] Granger Morgan and Max Henrion, "The Value of Knowing How Little You Know," Uncertainty: A Guide to Dealing with Uncertainty in Quantitative Risk and Policy Analysis, Cambridge University Press, 1990: 307.

[18] Presidential/Congressional Commission on Risk Assessment and Risk Management, Risk Commission Report, 1997, Volume 1: 39, Volume 2: 91.

[19] William W. Lowrance, Modern Science and Human Values, Oxford University Press, New York, NY 1985: 86.

Balance. While expertise is the primary consideration, reviewers should also be selected to represent a diversity of scientific perspectives relevant to the subject. On most controversial issues, there exists a range of respected scientific viewpoints regarding interpretation of the available literature. Inviting reviewers with competing views on the science may lead to a sharper, more focused peer review. Indeed, as a final layer of review, some organizations (e.g., the National Academy of Sciences) specifically recruit reviewers with strong opinions to test the scientific strength and balance of their reports. The NAS policy on committee composition and balance[20] highlights important considerations associated with perspective, bias, and objectivity.

Independence. In its narrowest sense, independence in a reviewer means that the reviewer was not involved in producing the draft document to be reviewed. However, for peer review of some documents, a broader view of independence is necessary to assure credibility of the process. Reviewers are generally not employed by the agency or office producing the document. As the National Academy of Sciences has stated, "external experts often can be more open, frank, and challenging to the status quo than internal reviewers, who may feel constrained by organizational concerns."[21] The Carnegie Commission on Science, Technology, and Government notes that "external science advisory boards serve a critically important function in providing regulatory agencies with expert advice on a range of issues."[22] However, the choice of reviewers requires a case-by-case analysis. Reviewers employed by other federal and state agencies may possess unique or indispensable expertise.

A related issue is whether government-funded scientists in universities and consulting firms have sufficient independence from the federal agencies that support their work to

---

[20] National Academy of Sciences, "Policy and Procedures on Committee Composition and Balance and Conflicts of Interest for Committees Used in the Development of Reports," May 2003: Available at: http://www.nationalacademies.org/coi/index.html.

[21] National Research Council, Peer Review in Environmental Technology Development Programs: The Department of Energy's Office of Science and Technology, National Academy Press, Washington, D.C., 1998: 3.

[22] Carnegie Commission on Science, Technology, and Government, Risk and the Environment. Improving Regulatory Decision Making, Carnegie Commission, New York, 1993: 90.

be appropriate peer reviewers for those agencies.[23] This concern can be mitigated in situations where the scientist initiates the hypothesis to be tested or the method to be developed, which effectively creates a buffer between the scientist and the agency. When an agency awards grants through a competitive process that includes peer review, the agency's potential to influence the scientist's research is limited. As such, when a scientist is awarded a government research grant through an investigator-initiated, peer- reviewed competition, there generally should be no question as to that scientist's ability to offer independent scientific advice to the agency on other projects. This contrasts, for example, to a situation in which a scientist has a consulting or contractual arrangement with the agency or office sponsoring a peer review. Likewise, when the agency and a researcher work together (e.g., through a cooperative agreement) to design or implement a study, there is less independence from the agency. Furthermore, if a scientist has repeatedly served as a reviewer for the same agency, some may question whether that scientist is sufficiently independent from the agency to be employed as a peer reviewer on agency-sponsored projects.

As the foregoing suggests, independence poses a complex set of questions that must be considered by agencies when peer reviewers are selected. In general, agencies shall make an effort to rotate peer review responsibilities across the available pool of qualified reviewers, recognizing that in some cases repeated service by the same reviewer is needed because of essential expertise.

Some agencies have built entire organizations to provide independent scientific advice while other agencies tend to employ ad hoc scientific panels on specific issues. Respect for the independence of reviewers may be enhanced if an agency collects names of potential reviewers (based on considerations of expertise and reputation for objectivity) from the public, including scientific or professional societies. The Department of Energy's use of the American Society of Mechanical Engineers to identify potential peer reviewers from a variety of different scientific societies provides an example of how

---

[23] Lars Noah, "Scientific 'Republicanism': Expert Peer Review and the Quest for Regulatory Deliberation, Emory Law Journal, Atlanta, Fall 2000:1066.

professional societies can assist in the development of an independent peer review panel.[24]

Conflict of Interest. The National Academy of Sciences defines "conflict of interest" as any financial or other interest that conflicts with the service of an individual on the review panel because it could impair the individual's objectivity or could create an unfair competitive advantage for a person or organization.[25] This standard provides a useful benchmark for agencies to consider in selecting peer reviewers. Agencies shall make a special effort to examine prospective reviewers' potential financial conflicts, including significant investments, consulting arrangements, employer affiliations and grants/contracts. Financial ties of potential reviewers to regulated entities (e.g., businesses), other stakeholders, and regulatory agencies shall be scrutinized when the information being reviewed is likely to be relevant to regulatory policy. The inquiry into potential conflicts goes beyond financial investments and business relationships and includes work as an expert witness, consulting arrangements, honoraria and sources of grants and contracts. To evaluate any real or perceived conflicts of interest with potential reviewers and questions regarding the independence of reviewers, agencies are referred to federal ethics requirements, applicable standards issued by the Office of Government Ethics, and the prevailing practices of the National Academy of Sciences. Specifically, peer reviewers who are federal employees (including special government employees) are subject to federal requirements governing conflicts of interest. See, e.g., 18 U.S.C. § 208; 5 C.F.R. Part 2635 (2004). With respect to reviewers who are not federal employees, agencies shall adopt or adapt the NAS policy for committee selection with respect to evaluating conflicts of interest.[26] Both the NAS and the federal government recognize that under certain circumstances some conflict may be unavoidable in order to obtain the necessary expertise. See, e.g., 18 U.S.C. § 208(b)(3); 5 U.S.C. App. § 15 (governing NAS committees). To improve the transparency of the process, when an agency

---

[24] American Society for Mechanical Engineers, Assessment of Technologies Supported by the Office of Science and Technology, Department of Energy: Results of the Peer Review for Fiscal Year 2002, ASME Technical Publishing, Danvers, MA, 2003.

[25] National Academy of Sciences, "Policy and Procedures on Committee Composition and Balance and Conflicts of Interest for Committees Used in the Development of Reports," May 2003: Available at: http://www.nationalacademies.org/coi/index.html.

determines that it is necessary to use a reviewer with a real or perceived conflict of interest, the agency should consider publicly disclosing those conflicts. In such situations, the agency shall inform potential reviewers of such disclosure at the time they are recruited.

Disclosure and Attribution: Anonymous versus Identified

Peer reviewers must have a clear understanding of how their comments will be conveyed to the authors of the document and to the public. When peer review of government reports is considered, the case for transparency is stronger, particularly when the report addresses an issue with significant ramifications for the public and private sectors. The public may not have confidence in the peer review process when the names and affiliations of the peer reviewers are unknown. Without access to the comments of reviewers, the public is incapable of determining whether the government has seriously considered the comments of reviewers and made appropriate revisions. Disclosure of the slate of reviewers and the substance of their comments can strengthen public confidence in the peer review process. It is common at many journals and research funding agencies to disclose annually the slate of reviewers. Moreover, the National Academy of Sciences now discloses the names of its peer reviewers, without disclosing the substance of their comments. The science advisory committees to regulatory agencies typically disclose at least a summary of the comments of reviewers as well as their names and affiliations.

For agency-sponsored peer review conducted under Sections II and III, this Bulletin strikes a compromise by requiring disclosure of the identity of the reviewers, but not public attribution of specific comments to specific reviewers. The agency has considerable discretion in the implementation of this compromise (e.g., summarizing the

---

[26] Ibid.

views of reviewers as a group or disclosing individual reviewer comments without attribution). Whatever approach is employed, the agency must inform reviewers in advance of how it intends to address this issue. Information about a reviewer retrieved from a record filed by the reviewer's name or other identifier may be disclosed only as permitted by the conditions of disclosure enumerated in the Privacy Act, 5 U.S.C. § 552a as amended, and as interpreted in OMB implementing guidance, 40 Fed. Reg. 28,948 (July 9, 1975).

Public Participation

Public comments can be important in shaping expert deliberations. Agencies may decide that peer review should precede an opportunity for public comment to ensure that the public receives the most scientifically strong product (rather than one that may change substantially as a result of peer reviewer suggestions). However, there are situations in which public participation in peer review is an important aspect of obtaining a high- quality product through a credible process. Agencies, however, should avoid open- ended comment periods, which may delay completion of peer reviews and complicate the completion of the final work product.

Public participation can take a variety of forms, including opportunities to provide oral comments before a peer review panel or requests to provide written comments to the peer reviewers. Another option is for agencies to publish a "request for comment" or other notice in which they solicit public comment before a panel of peer reviewers performs its work.

Disposition of Reviewer Comments

A peer review is considered completed once the agency considers and addresses the reviewers' comments. All reviewer comments should be given consideration and be incorporated where relevant and valid. For instance, in the context of risk assessments, the National Academy of Sciences recommends that peer review include a written evaluation made available for public inspection.[27] In cases where there is a public panel,

---

[27] National Research Council, Risk Assessment in the Federal Government: Managing the Process, National Academy Press, Washington, D.C., 1983.

the agency should plan publication of the peer review report(s) and the agency's response to peer reviewer comments.

In addition, the credibility of the final scientific report is likely to be enhanced if the public understands how the agency addressed the specific concerns raised by the peer reviewers. Accordingly, agencies should consider preparing a written response to the peer review report explaining: the agency's agreement or disagreement, the actions the agency has undertaken or will undertake in response to the report, and (if applicable) the reasons the agency believes those actions satisfy any key concerns or recommendations in the report.

Adequacy of Prior Peer Review

In light of the broad range of information covered by Section II, agencies are directed to choose a peer review mechanism that is adequate, giving due consideration to the novelty and complexity of the science to be reviewed, the relevance of the information to decision making, the extent of prior peer reviews, and the expected benefits and costs of additional review.

Publication in a refereed scientific journal may mean that adequate peer review has been performed. However, the intensity of peer review is highly variable across journals. There will be cases in which an agency determines that a more rigorous or transparent review process is necessary. For instance, an agency may determine a particular journal review process did not address questions (e.g., the extent of uncertainty inherent in a finding) that the agency determines should be addressed before disseminating that information. As such, prior peer review and publication is not by itself sufficient grounds for determining that no further review is necessary.

Section III: Peer Review of Highly Influential Scientific Assessments

Whereas Section II leaves most of the considerations regarding the form of the peer review to the agency's discretion, Section III requires a more rigorous form of peer review for highly influential scientific assessments. The requirements of Section II of this Bulletin apply to Section III, but Section III has some additional requirements, which are discussed below. In planning a peer review under Section III, agencies typically will have to devote greater resources and attention to the issues discussed in Section II, i.e., individual versus panel review; timing; scope of the review; selection of reviewers; disclosure and attribution; public participation; and disposition of reviewer comments.

A scientific assessment is considered "highly influential" if the agency or the OIRA Administrator determines that the dissemination could have a potential impact of more than $500 million in any one year on either the public or private sector or that the dissemination is novel, controversial, or precedent-setting, or has significant interagency interest. One of the ways information can exert economic impact is through the costs or benefits of a regulation based on the disseminated information. The qualitative aspect of this definition may be most useful in cases where it is difficult for an agency to predict the potential economic effect of dissemination. In the context of this Bulletin, it may be either the approach used in the assessment or the interpretation of the information itself that is novel or precedent-setting. Peer review can be valuable in establishing the bounds of the scientific debate when methods or interpretations are a source of controversy among interested parties. If information is covered by Section III, an agency is required to adhere to the peer review procedures specified in Section III.

Section III (2) clarifies that the principal findings, conclusions and recommendations in official reports of the National Academy of Sciences that fall under this Section are generally presumed not to require additional peer review. All other highly influential scientific assessments require a review that meets the requirements of Section III of this Bulletin.

With regard to the selection of reviewers, Section III(3)(a) emphasizes consideration of expertise and balance. As discussed in Section II, expertise refers to the required knowledge, experience and skills required to perform the review whereas balance refers to the need for diversity in scientific perspective and disciplines. We emphasize that the term "balance" here refers not to balancing of stakeholder or political interests but rather to a broad and diverse representation of respected perspectives and intellectual traditions within the scientific community, as discussed in the NAS policy on committee composition and balance.[28]

Section III (3)(b) instructs agencies to consider barring participation by scientists with a conflict of interest. The conflict of interest standards for Sections II and III of the Bulletin are identical. As discussed under Section II, those peer reviewers who are federal employees, including Special Government Employees, are subject to applicable statutory and regulatory standards for federal employees. For non-government employees, agencies shall adopt or adapt the NAS policy for committee member selection with respect to evaluating conflicts of interest.

Section III (3)(c) instructs agencies to ensure that reviewers are independent of the agency sponsoring the review. Scientists employed by the sponsoring agency are not permitted to serve as reviewers for highly influential scientific assessments. This does not preclude Special Government Employees, such as academics appointed to advisory committees, from serving as peer reviewers. The only exception to this ban would be the rare situation in which a scientist from a different agency of a Cabinet-level department than the agency that is disseminating the scientific assessment has expertise, experience and skills that are essential but cannot be obtained elsewhere. In evaluating the need for this exception, agencies shall use the NAS criteria for assessing the appropriateness of using employees of sponsors (e.g., the government scientist must not have had any part in the development or prior review of the scientific information and must not hold a position of managerial or policy responsibility).

---

[28] National Academy of Sciences, "Policy and Procedures on Committee Composition and Balance and Conflicts of Interest for Committees Used in the Development of Reports," May 2003: Available at: http://www.nationalacademies.org/coi/index.html.

We also considered whether a reviewer can be independent of the agency if that reviewer receives a substantial amount of research funding from the agency sponsoring the review. Research grants that were awarded to the scientist based on investigator-initiated, competitive, peer-reviewed proposals do not generally raise issues of independence. However, significant consulting and contractual relationships with the agency may raise issues of independence or conflict, depending upon the situation.

Section III (3)(d) addresses concerns regarding repeated use of the same reviewer in multiple assessments. Such repeated use should be avoided unless a particular reviewer's expertise is essential. Agencies should rotate membership across the available pool of qualified reviewers. Similarly, when using standing panels of scientific advisors, it is suggested that the agency rotate membership among qualified scientists in order to obtain fresh perspectives and reinforce the reality and perception of independence from the agency.

Section III (4) requires agencies to provide reviewers with sufficient background information, including access to key studies, data and models, to perform their role as peer reviewers. In this respect, the peer review envisioned in Section III is more rigorous than some forms of journal peer review, where the reviewer is often not provided access to underlying data or models. Reviewers shall be informed of applicable access, objectivity, reproducibility and other quality standards under federal information quality laws.

Section III (5) addresses opportunity for public participation in peer review, and provides that the agency shall, wherever possible, provide for public participation. In some cases, an assessment may be so sensitive that it is critical that the agency's assessment achieve a high level of quality before it is publicized. In those situations, a rigorous yet confidential peer review process may be appropriate, prior to public release of the assessment. If an agency decides to make a draft assessment publicly available at the onset of a peer review process, the agency shall, whenever possible, provide a vehicle for the public to provide written comments, make an oral presentation before the peer reviewers, or both. When written public comments are received, the agency shall ensure that peer reviewers receive copies of comments that address significant scientific issues with ample time to consider them in their review. To avoid undue delay of agency activities, the agency shall specify time limits for public participation throughout the peer review process.

Section III (6) requires that agencies instruct reviewers to prepare a peer review report that describes the nature and scope of their review and their findings and conclusions. The report shall disclose the name of each peer reviewer and a brief description of his or her organizational affiliation, credentials and relevant experiences. The peer review report should either summarize the views of the group as a whole (including any dissenting views) or include a verbatim copy of the comments of the individual reviewers (with or without attribution of specific views to specific names). The agency shall also prepare a written response to the peer review report, indicating whether the agency agrees with the reviewers and what actions the agency has taken or plans to take to address the points made by reviewers. The agency is required to disseminate the peer review report and the agency's response to the report on the agency's website, including all the materials related to the peer review such as the charge statement, peer review report, and agency response to the review. If the scientific information is used to support a final rule then, where practicable, the peer review report shall be made available to the public with enough time for the public to consider the implications of the peer review report for the rule being considered.

Section III (7) authorizes but does not require an agency to commission an entity independent of the agency to select peer reviewers and/or manage the peer review process in accordance with this Bulletin. The entity may be a scientific or professional society, a firm specializing in peer review, or a non-profit organization with experience in peer review.

Section IV: Alternative Procedures

Peer review as described in this Bulletin is only one of many procedures that agencies can employ to ensure an appropriate degree of pre-dissemination quality of influential scientific information. For example, Congress has assigned the NAS a special role in advising the federal government on scientific and technical issues. The procedures of the NAS are generally quite rigorous, and thus agencies should presume that major findings, conclusions, and recommendations of NAS reports meet the performance standards of this Bulletin.

As an alternative to complying with Sections II and III of this Bulletin, an agency may instead (1) rely on scientific information produced by the National Academy of Sciences, (2) commission the National Academy of Sciences to peer review an agency draft scientific information product, or (3) employ an alternative procedure or set of procedures, specifically approved by the OIRA Administrator in consultation with the Office of Science and Technology Policy (OSTP), that ensures that the scientific information product meets applicable information-quality standards.

An example of an alternative procedure is to commission a respected third party other than the NAS (e.g., the Health Effects Institute or the National Commission on Radiation Protection and Measurement) to conduct an assessment or series of related assessments. Another example of an alternative set of procedures is the three-part process used by the National Institutes of Health (NIH) to generate scientific guidance. Under that process, a scientific proposal or white paper is generated by a working group composed of external, independent scientific experts; that paper is then forwarded to a separate external scientific council, which then makes recommendations to the agency. The agency, in turn, decides whether to adopt and/or modify the proposal. For large science agencies that have diverse research portfolios and do not have significant regulatory responsibilities, such as NIH, an acceptable alternative would be to allow scientists from one part of the agency (for example, an NIH institute) to participate in the review of documents prepared by another part of the agency, as long as the head of the agency confirms in writing that each of the reviewers meets the NAS criteria relating to the appropriateness of using employees of sponsors (e.g., the government scientist must not have had any part in the development or prior review of the scientific information and must not hold a position of managerial or policy responsibility). The purpose of Section IV is to encourage these types of innovation in the methods used to ensure pre- dissemination quality control of influential scientific information.

The mere existence of a public comment process (e.g., notice-and-comment procedures under the Administrative Procedure Act) does not constitute adequate peer review or an "alternative process," because it does not assure that qualified, impartial specialists in relevant fields have performed a critical evaluation of the agency's draft product.[29]

Section V: Peer Review Planning

Section V requires agencies to begin a systematic process of peer review planning for influential scientific information (including highly influential scientific assessments) that the agency plans to disseminate in the foreseeable future. A key feature of this planning process is a web-accessible listing of forthcoming influential scientific disseminations (i.e., an agenda) that is regularly updated by the agency. By making these plans publicly available, agencies will be able to gauge the extent of public interest in the peer review process for influential scientific information, including highly influential scientific assessments. These web-accessible agendas can also be used by the public to monitor agency compliance with this Bulletin.

Each entry on the agenda shall include a preliminary title of the planned report, a short paragraph describing the subject and purpose of the planned report, and an agency contact person. The agency shall provide its prediction regarding whether the dissemination will be "influential scientific information" or a "highly influential scientific assessment," as the designation can influence the type of peer review to be undertaken.

---

[29] William W. Lowrance, Modern Science and Human Values, Oxford University Press, New York, NY 1985: 86.

The agency shall discuss the timing of the peer review, as well as the use of any deferrals. Agencies shall include entries in the agenda for influential scientific information, including highly influential scientific assessments, for which the Bulletin's requirements have been deferred or waived. If the agency, in consultation with the OIRA Administrator, has determined that it is appropriate to use a Section IV "alternative procedure" for a specific dissemination, a description of that alternative procedure shall be included in the agenda.

Furthermore, for each entry on the agenda, the agency shall describe the peer review plan. Each peer review plan shall include: (i) a paragraph including the title, subject and purpose of the planned report, as well as an agency contact to whom inquiries may be directed to learn the specifics of the plan; (ii) whether the dissemination is likely to be influential scientific information or a highly influential scientific assessment; (iii) the timing of the review (including deferrals); (iv) whether the review will be conducted through a panel or individual letters (or whether an alternative procedure will be exercised); (v) whether there will be opportunities for the public to comment on the work product to be peer reviewed, and if so, how and when these opportunities will be provided; (vi) whether the agency will provide significant and relevant public comments to the peer reviewers before they conduct their review; (vii) the anticipated number of reviewers (3 or fewer; 4-10; or more than 10); (viii) a succinct description of the primary disciplines or expertise needed in the review; (ix) whether reviewers will be selected by the agency or by a designated outside organization; and (x) whether the public, including scientific or professional societies, will be asked to nominate potential peer reviewers.

The agency shall provide a link from the agenda to each document made public pursuant to this Bulletin. Agencies shall link their peer review agendas to the U.S. Government's official web portal: firstgov at http://www.FirstGov.gov

Agencies should update their peer review agendas at least every six months. However, in some cases -- particularly for highly influential scientific assessments and other particularly important information -- more frequent updates of existing entries on the agenda, or the addition of new entries to the agenda, may be warranted. When new entries are added to the agenda of forthcoming reports and other information, the public should be provided with sufficient time to comment on the agency's peer review plan for that report or product. Agencies shall consider public comments on the peer review plan. Agencies are encouraged to offer a listserve or similar mechanism for members of the public who would like to be notified by email each time an agency's peer review agenda has been updated.

The peer review planning requirements of this Bulletin are designed to be implemented in phases. Specifically, the planning requirements of the Bulletin will go into effect for documents subject to Section III of the Bulletin (highly influential scientific assessments) six months after publication. However, the planning requirements for documents subject to Section II of the Bulletin do not go into effect until one year after publication. It is expected that agency experience with the planning requirements of the Bulletin for the smaller scope of documents encompassed in Section III will be used to inform implementation of these planning requirements for the larger scope of documents covered under Section II.

Section VI: Annual Report

Each agency shall prepare an annual report that summarizes key decisions made pursuant to this Bulletin. In particular, each agency should provide to OIRA the following: 1) the number of peer reviews conducted subject to the Bulletin (i.e., for influential scientific information and highly influential scientific assessments); 2) the number of times alternative procedures were invoked; 3) the number of times waivers or deferrals were invoked (and in the case of deferrals, the length of time elapsed between the deferral and the peer review); 4) any decision to appoint a reviewer pursuant to any exception to the applicable independence or conflict of interest standards of the Bulletin, including determinations by the Secretary or Deputy Secretary pursuant to Section III (3) (c); 5) the number of peer review panels that were conducted in public and the number that allowed public comment; 6) the number of public comments provided on the agency's peer review plans; and 7) the number of peer reviewers that the agency used that were recommended by professional societies.

Section VII: Certification in the Administrative Record

If an agency relies on influential scientific information or a highly influential scientific assessment subject to the requirements of this Bulletin in support of a regulatory action, the agency shall include in the administrative record for that action a certification that explains how the agency has complied with the requirements of this Bulletin and the Information Quality Act. Relevant materials are to be placed in the administrative record.

Section VIII: Safeguards, Deferrals, and Waivers

Section VIII recognizes that individuals serving as peer reviewers have a privacy interest in information about themselves that the government maintains and retrieves by name or identifier from a system of records. To the extent information about a reviewer (name, credential, affiliation) will be disclosed along with his/her comments or analysis, the agency must comply with the requirements of the Privacy Act, 5 U.S.C. 552a, as amended, and OMB Circular A-130, Appendix I, 61 Fed. Reg. 6428 (February 20, 1996) to establish appropriate routine uses in a published System of Records Notice. Furthermore, the peer review must be conducted in a manner that respects confidential business information as well as intellectual property.

Section VIII also allows for a deferral or waiver of the requirements of the Bulletin where necessary. Specifically, the agency head may waive or defer some or all of the peer review requirements of Sections II or III of this Bulletin if there is a compelling rationale for waiver or deferral. Waivers will seldom be warranted under this provision because the Bulletin already provides significant safety valves, such as: the exemptions provided in Section IX, including the exemption for time-sensitive health and safety information;

the authorization for alternative procedures in Section IV; and the overall flexibility provided for peer reviews of influential scientific information under Section II. Nonetheless, we have included this waiver and deferral provision to ensure needed flexibility in unusual and compelling situations not otherwise covered by the exemptions to the Bulletin, such as situations where unavoidable legal deadlines prevent full compliance with the Bulletin before information is disseminated. Deadlines found in consent decrees agreed to by agencies after the Bulletin is issued will not ordinarily warrant waiver of the Bulletin's requirements because those deadlines should be negotiated to permit time for all required procedures, including peer review. In addition, when an agency is unavoidably up against a deadline, deferral of some or all requirements of the Bulletin (as opposed to outright waiver of all of them) is the most appropriate accommodation between the need to satisfy immovable deadlines and the need to undertake proper peer review. If the agency head defers any of the peer review requirements prior to dissemination, peer review should be conducted as soon as practicable thereafter.

Section IX: Exemptions

There are a variety of situations where agencies need not conduct peer review under this Bulletin. These include, for example, disseminations of sensitive information related to certain national security, foreign affairs, or negotiations involving international treaties and trade where compliance with this Bulletin would interfere with the need for secrecy or promptness.

This Bulletin does not cover official disseminations that arise in adjudications and permit proceedings, unless the agency determines that peer review is practical and appropriate and that the influential dissemination is scientifically or technically novel (i.e., a major change in accepted practice) or likely to have precedent-setting influence on future adjudications or permit proceedings. This exclusion is intended to cover, among other things, licensing, approval and registration processes for specific product development activities as well as site-specific activities. The determination as to whether peer review is practical and appropriate is left to the discretion of the agency. While this Bulletin is not broadly applicable to adjudications, agencies are encouraged to hold peer reviews of scientific assessments supporting adjudications to the same technical standards as peer reviews covered by the Bulletin, including transparency and disclosure of the data and models underlying the assessments. Protections apply to confidential business information.

The Bulletin does not cover time-sensitive health and safety disseminations, for example, a dissemination based primarily on data from a recent clinical trial that was adequately peer reviewed before the trial began. For this purpose, "health" includes public health, or plant or animal infectious diseases.

This Bulletin covers original data and formal analytic models used by agencies in Regulatory Impact Analyses (RIAs). However, the RIA documents themselves are already reviewed through an interagency review process under E.O. 12866 that involves application of the principles and methods defined in OMB Circular A-4. In that respect, RIAs are excluded from coverage by this Bulletin, although agencies are encouraged to have RIAs reviewed by peers within the government for adequacy and completeness.

The Bulletin does not cover accounting, budget, actuarial, and financial information including that which is generated or used by agencies that focus on interest rates, banking, currency, securities, commodities, futures, or taxes.

Routine statistical information released by federal statistical agencies (e.g., periodic demographic and economic statistics) and analyses of these data to compute standard indicators and trends (e.g., unemployment and poverty rates) is excluded from this Bulletin.

The Bulletin does not cover information disseminated in connection with routine rules that materially alter entitlements, grants, user fees, or loan programs, or the rights and obligations of recipients thereof.

If information is disseminated pursuant to an exemption to this Bulletin, subsequent disseminations are not automatically exempted. For example, if influential scientific information is first disseminated in the course of an exempt agency adjudication, but is later disseminated in the context of a non-exempt rulemaking, the subsequent dissemination will be subject to the requirements of this Bulletin even though the first dissemination was not.

Section X: OIRA and OSTP Responsibilities

OIRA, in consultation with OSTP, is responsible for overseeing agency implementation of this Bulletin. In order to foster learning about peer review practices across agencies, OIRA and OSTP shall form an interagency workgroup on peer review that meets regularly, discusses progress and challenges, and recommends improvements to peer review practices.

Section XI: Effective Date and Existing Law

The requirements of this Bulletin, with the exception of Section V, apply to information disseminated on or after six months after publication of this Bulletin. However, the Bulletin does not apply to information that is already being addressed by an agency- initiated peer review process (e.g., a draft is already being reviewed by a formal scientific advisory committee established by the agency). An existing peer review mechanism mandated by law should be implemented by the agency in a manner as consistent as possible with the practices and procedures outlined in this Bulletin. The requirements of Section V apply to "highly influential scientific assessments," as designated in Section III of the Bulletin, within six months of publication of the final Bulletin. The requirements in Section V apply to documents subject to Section II of the Bulletin one year after publication of the final Bulletin.

Section XII: Judicial Review

This Bulletin is intended to improve the internal management of the Executive Branch and is not intended to, and does not, create any right or benefit, substantive or procedural, enforceable at law or in equity, against the United States, its agencies or other entities, its officers or employees, or any other person.

Bulletin for Peer Review

I. Definitions.

For purposes of this Bulletin --

1. the term "Administrator" means the Administrator of the Office of Information and Regulatory Affairs in the Office of Management and Budget (OIRA);

2. the term "agency" has the same meaning as in the Paperwork Reduction Act, 44 U.S.C. § 3502(1);

3. the term "dissemination" means agency initiated or sponsored distribution of information to the public (see 5 C.F.R. 1320.3(d) (definition of "Conduct or Sponsor")). Dissemination does not include distribution limited to government employees or agency contractors or grantees; intra- or inter-agency use or sharing of government information; or responses to requests for agency records under the Freedom of Information Act, the Privacy Act, the Federal Advisory Committee Act, the Government Performance and Results Act or similar law. This definition also excludes distribution limited to correspondence with individuals or persons, press releases, archival records, public filings, subpoenas and adjudicative processes. The term "dissemination" also excludes information distributed for peer review in compliance with this Bulletin, provided that the distributing agency includes a clear disclaimer on the information as follows: "THIS INFORMATION IS DISTRIBUTED SOLELY FOR THE PURPOSE OF PRE-DISSEMINATION PEER REVIEW UNDER APPLICABLE INFORMATION

QUALITY GUIDELINES. IT HAS NOT BEEN FORMALLY DISSEMINATED BY [THE AGENCY]. IT DOES NOT REPRESENT AND SHOULD NOT BE CONSTRUED TO REPRESENT ANY AGENCY DETERMINATION OR POLICY." For the purposes of this Bulletin, "dissemination" excludes research produced by government-funded scientists (e.g., those supported extramurally or intramurally by federal agencies or those working in state or local governments with federal support) if that information does not represent the views of an agency. To qualify for this exemption, the information should display a clear disclaimer that "the findings and conclusions in this report are those of the author(s) and do not necessarily represent the views of the funding agency";

4. the term "Information Quality Act" means Section 515 of Public Law 106-554 (Pub. L. No. 106-554, § 515, 114 Stat. 2763, 2763A-153-154 (2000));

5. the term "scientific information" means factual inputs, data, models, analyses, technical information, or scientific assessments based on the behavioral and social sciences, public health and medical sciences, life and earth sciences, engineering, or physical sciences. This includes any communication or representation of knowledge such as facts or data, in any medium or form, including textual, numerical, graphic, cartographic, narrative, or audiovisual forms. This definition includes information that an agency disseminates from a web page, but does not include the provision of hyperlinks to information that others disseminate. This definition does not include opinions, where the agency's presentation makes clear that what is being offered is someone's opinion rather than fact or the agency's views;

6. the term "influential scientific information" means scientific information the agency reasonably can determine will have or does have a clear and substantial impact on important public policies or private sector decisions; and

7. the term "scientific assessment" means an evaluation of a body of scientific or technical knowledge, which typically synthesizes multiple factual inputs, data, models, assumptions, and/or applies best professional judgment to bridge uncertainties in the available information. These assessments include, but are not limited to, state-of-science reports; technology assessments; weight-of-evidence analyses; meta-analyses; health,

safety, or ecological risk assessments; toxicological characterizations of substances; integrated assessment models; hazard determinations; or exposure assessments.

II. Peer Review of Influential Scientific Information.

1. In General: To the extent permitted by law, each agency shall conduct a peer review on all influential scientific information that the agency intends to disseminate. Peer reviewers shall be charged with reviewing scientific and technical matters, leaving policy determinations for the agency. Reviewers shall be informed of applicable access, objectivity, reproducibility and other quality standards under the federal laws governing information access and quality.

2. Adequacy of Prior Peer Review: For information subject to this section of the Bulletin, agencies need not have further peer review conducted on information that has already been subjected to adequate peer review. In determining whether prior peer review is adequate, agencies shall give due consideration to the novelty and complexity of the science to be reviewed, the importance of the information to decision making, the extent of prior peer reviews, and the expected benefits and costs of additional review. Principal findings, conclusions and recommendations in official reports of the National Academy of Sciences are generally presumed to have been adequately peer reviewed.

3. Selection of Reviewers:

    (a) Expertise and Balance: Peer reviewers shall be selected based on expertise, experience and skills, including specialists from multiple disciplines, as necessary. The group of reviewers shall be sufficiently broad and diverse to fairly represent the relevant scientific and technical perspectives and fields of knowledge. Agencies shall consider requesting that the public, including scientific and professional societies, nominate potential reviewers.

    (b) Conflicts: The agency – or the entity selecting the peer reviewers – shall (i) ensure that those reviewers serving as federal employees (including special government employees) comply with applicable federal ethics requirements; (ii) in selecting peer reviewers who are not government employees, adopt or adapt the National Academy of Sciences policy for committee selection with respect to evaluating the potential for conflicts (e.g., those arising from investments; agency, employer, and business affiliations; grants, contracts and consulting income). For scientific information relevant to specific regulations, the agency shall examine a reviewer's financial ties to regulated entities (e.g., businesses), other stakeholders, and the agency.

(c) Independence: Peer reviewers shall not have participated in development of the work product. Agencies are encouraged to rotate membership on standing panels across the pool of qualified reviewers. Research grants that were awarded to scientists based on investigator-initiated, competitive, peer-reviewed proposals generally do not raise issues as to independence or conflicts.

4. Choice of Peer Review Mechanism: The choice of a peer review mechanism (for example, letter reviews or ad hoc panels) for influential scientific information shall be based on the novelty and complexity of the information to be reviewed, the importance of the information to decision making, the extent of prior peer review, and the expected benefits and costs of review, as well as the factors regarding transparency described in II(5).

5. Transparency: The agency -- or entity managing the peer review -- shall instruct peer reviewers to prepare a report that describes the nature of their review and their findings and conclusions. The peer review report shall either (a) include a verbatim copy of each reviewer's comments (either with or without specific attributions) or (b) represent the views of the group as a whole, including any disparate and dissenting views. The agency shall disclose the names of the reviewers and their organizational affiliations in the report. Reviewers shall be notified in advance regarding the extent of disclosure and attribution planned by the agency. The agency shall disseminate the final peer review report on the agency's website along with all materials related to the peer review (any charge statement, the peer review report, and any agency response). The peer review report shall be discussed in the preamble to any related rulemaking and included in the administrative record for any related agency action.

6. Management of Peer Review Process and Reviewer Selection: The agency may commission independent entities to manage the peer review process, including the selection of peer reviewers, in accordance with this Bulletin.

III. Additional Peer Review Requirements for Highly Influential Scientific Assessments.

1. Applicability: This section applies to influential scientific information that the agency or the Administrator determines to be a scientific assessment that:

    (i) could have a potential impact of more than $500 million in any year, or

    (ii) is novel, controversial, or precedent-setting or has significant interagency interest.

2. In General: To the extent permitted by law, each agency shall conduct peer reviews on all information subject to this section. The peer reviews shall satisfy the requirements of Section II of this Bulletin, as well as the additional requirements found in this section. Principal findings, conclusions and recommendations in official reports of the National Academy of Sciences that fall under this section are generally presumed not to require additional peer review.

3. Selection of Reviewers:

    (a) Expertise and Balance: Peer reviewers shall be selected based on expertise, experience and skills, including specialists from multiple disciplines, as necessary. The group of reviewers shall be sufficiently broad and diverse to fairly represent the relevant scientific and technical perspectives and fields of knowledge. Agencies shall consider requesting that the public, including scientific and professional societies, nominate potential reviewers.

    (b) Conflicts: The agency – or the entity selecting the peer reviewers – shall (i) ensure that those reviewers serving as federal employees (including special government employees) comply with applicable federal ethics requirements; (ii) in selecting peer reviewers who are not government employees, adopt or adapt the National Academy of Sciences' policy for committee selection with respect to evaluating the potential for conflicts (e.g., those arising from investments; agency, employer, and business affiliations; grants, contracts and consulting income). For scientific assessments relevant to specific regulations, a reviewer's financial ties to regulated entities (e.g., businesses), other stakeholders, and the agency shall be examined.

(c) Independence: In addition to the requirements of Section II (3)(c), which shall apply to all reviews conducted under Section III, the agency -- or entity selecting the reviewers -- shall bar participation of scientists employed by the sponsoring agency unless the reviewer is employed only for the purpose of conducting the peer review (i.e., special government employees). The only exception to this bar would be the rare case where the agency determines, using the criteria developed by NAS for evaluating use of "employees of sponsors," that a premier government scientist is (a) not in a position of management or policy responsibility and (b) possesses essential expertise that cannot be obtained elsewhere. Furthermore, to be eligible for this exception, the scientist must be employed by a different agency of the Cabinet-level department than the agency that is disseminating the scientific information. The agency's determination shall be documented in writing and approved, on a non-delegable basis, by the Secretary or Deputy Secretary of the department prior to the scientist's appointment.

(d) Rotation: Agencies shall avoid repeated use of the same reviewer on multiple assessments unless his or her participation is essential and cannot be obtained elsewhere.

4. Information Access: The agency -- or entity managing the peer review -- shall provide the reviewers with sufficient information -- including background information about key studies or models -- to enable them to understand the data, analytic procedures, and assumptions used to support the key findings or conclusions of the draft assessment.

5. Opportunity for Public Participation: Whenever feasible and appropriate, the agency shall make the draft scientific assessment available to the public for comment at the same time it is submitted for peer review (or during the peer review process) and sponsor a public meeting where oral presentations on scientific issues can be made to the peer reviewers by interested members of the public. When employing a public comment process as part of the peer review, the agency shall, whenever practical, provide peer reviewers with access to public comments that address significant scientific or technical issues. To ensure that public participation does not unduly delay agency activities, the agency shall clearly specify time limits for public participation throughout the peer review process.

6. Transparency: In addition to the requirements specified in II(5), which shall apply to all reviews conducted under Section III, the peer review report shall include the charge to the reviewers and a short paragraph on both the credentials and relevant experiences of each peer reviewer. The agency shall prepare a written response to the peer review report explaining (a) the agency's agreement or disagreement with the views expressed in the report, (b) the actions the agency has undertaken or will undertake in response to the report, and (c) the reasons the agency believes those actions satisfy the key concerns stated in the report (if applicable). The agency shall disseminate its response to the peer review report on the agency's website with the related material specified in Section II(5).

7. Management of Peer Review Process and Reviewer Selection: The agency may commission independent entities to manage the peer review process, including the selection of peer reviewers, in accordance with this Bulletin.

IV. Alternative Procedures.

As an alternative to complying with Sections II and III of this Bulletin, an agency may instead: (i) rely on the principal findings, conclusions and recommendations of a report produced by the National Academy of Sciences; (ii) commission the National Academy of Sciences to peer review an agency's draft scientific information; or (iii) employ an alternative scientific procedure or process, specifically approved by the Administrator in consultation with the Office of Science and Technology Policy (OSTP), that ensures the agency's scientific information satisfies applicable information quality standards. The alternative procedure(s) may be applied to a designated report or group of reports.

V. Peer Review Planning.

1. Peer Review Agenda: Each agency shall post on its website, and update at least every six months, an agenda of peer review plans. The agenda shall describe all planned and ongoing influential scientific information subject to this Bulletin. The agency shall provide a link from the agenda to each document that has been made public pursuant to

this Bulletin. Agencies are encouraged to offer a listserve or similar mechanism to alert interested members of the public when entries are added or updated.

2. Peer Review Plans: For each entry on the agenda the agency shall describe the peer review plan. Each peer review plan shall include: (i) a paragraph including the title, subject and purpose of the planned report, as well as an agency contact to whom inquiries may be directed to learn the specifics of the plan; (ii) whether the dissemination is likely to be influential scientific information or a highly influential scientific assessment; (iii) the timing of the review (including deferrals); (iv) whether the review will be conducted through a panel or individual letters (or whether an alternative procedure will be employed); (v) whether there will be opportunities for the public to comment on the work product to be peer reviewed, and if so, how and when these opportunities will be provided; (vi) whether the agency will provide significant and relevant public comments to the peer reviewers before they conduct their review; (vii) the anticipated number of reviewers (3 or fewer; 4-10; or more than 10); (viii) a succinct description of the primary

disciplines or expertise needed in the review; (ix) whether reviewers will be selected by the agency or by a designated outside organization; and (x) whether the public, including scientific or professional societies, will be asked to nominate potential peer reviewers.

3. Public Comment: Agencies shall establish a mechanism for allowing the public to comment on the adequacy of the peer review plans. Agencies shall consider public comments on peer review plans.

VI. Annual Reports.

Each agency shall provide to OIRA, by December 15 of each year, a summary of the peer reviews conducted by the agency during the fiscal year. The report should include the following: 1) the number of peer reviews conducted subject to the Bulletin (i.e., for influential scientific information and highly influential scientific assessments); 2) the number of times alternative procedures were invoked; 3) the number of times waivers or deferrals were invoked (and in the case of deferrals, the length of time elapsed between the deferral and the peer review); 4) any decision to appoint a reviewer pursuant to any exception to the applicable independence or conflict of interest standards of the Bulletin, including determinations by the Secretary pursuant to Section III(3)(c); 5) the number of peer review panels that were conducted in public and the number that allowed public comment; 6) the number of public comments provided on the agency's peer review plans; and 7) the number of peer reviewers that the agency used that were recommended by professional societies.

VII. Certification in the Administrative Record.

If an agency relies on influential scientific information or a highly influential scientific assessment subject to this Bulletin to support a regulatory action, it shall include in the administrative record for that action a certification explaining how the agency has complied with the requirements of this Bulletin and the applicable information quality guidelines. Relevant materials shall be placed in the administrative record.

XIII. Safeguards, Deferrals, and Waivers.

1. Privacy: To the extent information about a reviewer (name, credentials, affiliation) will be disclosed along with his/her comments or analysis, the agency shall comply with the requirements of the Privacy Act, 5 U.S.C. § 522a as amended, and OMB Circular A-130, Appendix I, 61 Fed. Reg. 6428 (February 20, 1996) to establish appropriate routine uses in a published System of Records Notice.
2. Confidentiality: Peer review shall be conducted in a manner that respects (i) confidential business information and (ii) intellectual property.
3. Deferral and Waiver: The agency head may waive or defer some or all of the peer review requirements of Sections II and III of this Bulletin where warranted by a compelling rationale. If the agency head defers the peer review requirements prior to dissemination, peer review shall be conducted as soon as practicable.

IX. Exemptions.

Agencies need not have peer review conducted on information that is:

1. related to certain national security, foreign affairs, or negotiations involving international trade or treaties where compliance with this Bulletin would interfere with the need for secrecy or promptness;

2. disseminated in the course of an individual agency adjudication or permit proceeding (including a registration, approval, licensing, site-specific determination), unless the agency determines that peer review is practical and appropriate and that the influential dissemination is scientifically or technically novel or likely to have precedent-setting influence on future adjudications and/or permit proceedings;

3. a health or safety dissemination where the agency determines that the dissemination is time-sensitive (e.g., findings based primarily on data from a recent clinical trial that was adequately peer reviewed before the trial began);

4. an agency regulatory impact analysis or regulatory flexibility analysis subject to interagency review under Executive Order 12866, except for underlying data and analytical models used;

5. routine statistical information released by federal statistical agencies (e.g., periodic demographic and economic statistics) and analyses of these data to compute standard indicators and trends (e.g., unemployment and poverty rates);

6. accounting, budget, actuarial, and financial information, including that which is generated or used by agencies that focus on interest rates, banking, currency, securities, commodities, futures, or taxes; or

7. information disseminated in connection with routine rules that materially alter entitlements, grants, user fees, or loan programs, or the rights and obligations of recipients thereof.

X. Responsibilities of OIRA and OSTP.

OIRA, in consultation with OSTP, shall be responsible for overseeing implementation of this Bulletin. An interagency group, chaired by OSTP and OIRA, shall meet periodically to foster better understanding about peer review practices and to assess progress in implementing this Bulletin.

XI. Effective Date and Existing Law.

The requirements of this Bulletin, with the exception of those in Section V (Peer Review Planning), apply to information disseminated on or after six months following publication of this Bulletin, except that they do not apply to information for which an agency has already provided a draft report and an associated charge to peer reviewers. Any existing peer review mechanisms mandated by law shall be employed in a manner as consistent as possible with the practices and procedures laid out herein. The requirements in Section V apply to "highly influential scientific assessments," as designated in Section III of this Bulletin, within six months of publication of this Bulletin. The requirements in Section V apply to documents subject to Section II of this Bulletin one year after publication of this Bulletin.

XII. Judicial Review.

This Bulletin is intended to improve the internal management of the executive branch, and is not intended to, and does not, create any right or benefit, substantive or procedural, enforceable at law or in equity, against the United States, its agencies or other entities, its officers or employees, or any other person.

APPENDIX C. OVERVIEW OF THE AGENCY'S GENERAL ASSESSMENT FACTORS

In 2003, the Agency published, *A Summary of General Assessment Factors for Evaluating the Quality of Scientific and Technical Information*, in an effort to enhance the transparency about EPA's quality expectations for information that is voluntarily submitted to, or gathered, or generated, by the Agency for various purposes. The *Assessment Factors* document is intended to inform information-generating scientists about quality issues that should appropriately be taken into consideration at the time information is generated. It is also an additional resource for Agency staff as they evaluate the quality and relevance of information, regardless of source. The general assessment factors are drawn from the Agency's existing information quality systems, practices and guidelines that describe the types of considerations EPA takes into account when evaluating the quality and relevance of scientific and technical information used in support of Agency actions. The document is intended to raise the awareness of the information-generating public about EPA's ongoing interest in ensuring and enhancing the quality of information available for Agency use.

When evaluating the quality and relevance of scientific and technical information, the considerations that the Agency typically takes into account can be characterized by five general assessment factors:

- **Soundness** -The extent to which the scientific and technical procedures, measures, methods or models employed to generate the information are reasonable for, and consistent with, the intended application.

- **Applicability and Utility** -The extent to which the information is relevant for the Agency's intended use.

- **Clarity and Completeness** -The degree of clarity and completeness with which the data, assumptions, methods, quality assurance, sponsoring organizations and analyses employed to generate the information are documented.

- **Uncertainty and Variability** -The extent to which the variability and uncertainty (quantitative and qualitative) in the information or in the procedures, measures, methods or models are evaluated and characterized.

- **Evaluation and Review** -The extent of independent verification, validation and peer review of the information or of the procedures, measures, methods or models.

These assessment factors reflect the most salient features of EPA's existing information quality policies and guidelines.

For further information, please visit http://www2.epa.gov/osa/summary-general-assessment-factors-evaluating-quality-scientific-and-technical-information.

.

APPENDIX D. SOUND SCIENCE AND PEER REVIEW IN RULEMAKING POLICY

In response to several provisions of the December 2004, OMB Bulletin Final Information Quality Bulletin for Peer Review, the Office of Policy (formerly known as the Office of Policy, Economics, and Innovation [OPEI]) created conditional peer review template language for the preambles to proposed and final rules (Attachment A). This language should be used by rulewriters in the preamble of regulations that rely on influential scientific information or a highly influential scientific assessment, which are two categories of information defined in Section 3.2 of this Handbook.

For proposed and final regulations that rely on influential scientific information or a highly influential scientific assessment, rulewriters should use the template as a model to discuss peer review in the preamble where appropriate. In addition, peer review leaders should communicate with rulewriters and workgroup chairs to ensure that all appropriate peer review material is included in the docket, and that template language is included in the preamble.

The Office of Policy also revised the Action Memorandum Framework to include a discussion of peer review for influential scientific information or a highly influential scientific assessment (Attachment B).

ATTACHMENT A

Peer Review (Conditional Template)

Read this first (but <u>DO NOT</u> insert it in your preamble):

The OMB Final Information Quality Bulletin for Peer Review directs EPA to include a discussion of the peer review report and how the Agency complied with the provisions of the Bulletin in the preamble of rulemakings that are supported by influential scientific information or highly influential scientific assessments. Peer review reports should either (a) include a verbatim copy of each reviewer's comments (either with or without specific attributions) or (b) represent the views of the group as a whole, including any disparate and dissenting views. The Agency should disclose the names of the reviewers and their organizational affiliations in the report and should notify the reviewers in advance regarding the extent of the disclosure and attribution planned by the Agency. You should ensure that the peer review report is placed in the docket to comply with the OMB Bulletin.

Use this template if your proposed or final rule is based on a work product containing influential scientific information or a highly influential scientific assessment. This language should appear in the Supplementary Information section of regulatory preambles under General Information. You may want to include the language under the heading:

Did EPA conduct a peer review before issuing this notice?

Π **PROPOSED & FINAL ACTIONS: If you used a highly influential scientific assessment or influential scientific information to support this rulemaking, insert this into the preamble of your proposed or final rule, advanced notice of proposed rulemaking, or other substantive action:**

> This regulatory action was supported by **[influential scientific information or a highly influential scientific assessment]**. Therefore, EPA conducted a peer review in accordance with OMB's Final Information Quality Bulletin for Peer Review. **[Insert a brief description of the peer review process along with any other relevant information.]** The peer review report is located in the docket for this action. According to the report, **[insert a brief discussion of the peer review report. For more information about the peer review report, see the Peer Review Handbook.]**

Guidelines and Template for
Action Memoranda Accompanying Regulatory Packages
(Updated 10/05/2011)

Background

This guidance and template focuses only on those action memoranda prepared for the Administrator. You may or may not be asked by your program office to produce similar memos for actions signed by a delegated official other than the Administrator, but this guidance and template do not cover such instances. Speak with your program or regional office's Regulatory Steering Committee (RSC) member to learn what office-specific procedures may exist.

An action memorandum should be included with all regulatory packages brought to the Administrator for signature. Also, a copy of the draft memorandum should be included as part of 1) the Final Agency Review (FAR) package that is circulated to participating offices for final review and 2) the package that is submitted to the Office of Policy (OP) to initiate review by the Office of Management and Budget (OMB). The action memorandum provides a formal communication between the recommending official and the Administrator. It also offers a succinct rationale for the action, and provides a plain English explanation of the action in order to inform the Administrator's decision and help in future communications of the rule to the public and Congress. The memo should be signed by the recommending official [usually the Assistant Administrator (AA)] and should receive the personal attention of the recommending official.

Guidelines for Using This Template

Instructions for each section of an action memorandum are provided within the template below. The template is already formatted according to the *Correspondence Manual's* guidance, and you should be able to copy and paste the entire template into a new Microsoft Word document to begin creating your action memorandum. Please be aware that formatting may or may not be altered when you copy the template into another document. Carefully read the tips below to understand how your memo should be formatted.

As with other Action Development Process (ADP) guidance and templates, template text provided herein that appears in regular font and black ink should be inserted into your document without significant changes. Instructions on additional text to insert appear as bolded blue text in square brackets [like this]. Text that appears within blue curly brackets {like this} is optional to include and may be omitted without further consultation. Once you insert the appropriate text, please remove the brackets, instructions, color and unnecessary formatting from your document.

As indicated by the use of non-mandatory language such as "should," "recommend" and "may," this document provides recommendations and does not impose any legally binding requirements. Programs may include information on additional topics if they are relevant to a given action (e.g., information quality issues).

While preparing your action memorandum, you should follow these tips:

- Keep your memo to 4 pages; use attachments if you need to include longer descriptions.

- Use plain English. Provide a clear understanding of the action being taken and its impact; you should refrain from copying technical language from your action's preamble or regulatory text.

- In the header or footer of each page, insert this reminder: Internal, Deliberative Document – Do Not Cite, Quote or Distribute.

- Follow EPA's *Correspondence Manual* (http://intranet.epa.gov/agcyintr/manual/) guidance on:

  o Usage of "agency" and "EPA": Use "U.S. Environmental Protection Agency" as the first reference and "EPA" as the second reference. Use a lowercase "agency" in such references when it is not used as part of the full formal name of the agency.

  o Contractions: Do not use them.

  o Printing: Double-sided.

  o Ink: Print in black ink when printing. Signing officials may sign in blue or black ink. No other colors for official correspondence, aside from whatever colors may be on your letterhead.

  o Typeface:

    ▪ Font: 12 point, Times New Roman.

    ▪ Spacing:

      ➢ Lines should be single spaced, but leave one blank line between each paragraph. For headings, one blank line should be above the heading and zero blank lines should be below the heading.

      ➢ One space between each sentence and all punctuation marks.

    ▪ Indentation: Do not indent the first line of a paragraph.

  o Margins:

    ▪ 0.75-inch on all four sides.

    ▪ Align left for normal text. Do not center, justify or right-align. You may deviate from left justification when formatting bulleted or numbered lists, quotes or other special passages.

    ▪ Seven or eight hard returns should align the first line of text on the first page of the document, so that the first line is just below the office name that appears on the right-hand side of standard letterhead.

  o Page numbers:

    ▪ Use them only for memos consisting of four or more pages, and then begin numbering with page 3. For example, most action memos are likely to be four pages long. You would place page numbers on pages 3 and 4.

      ➢ Note: You will need to use a "section break" rather than a "page break" in Microsoft Word to accomplish this formatting. Also, you must remove the "Link to Previous Section" feature for the section in which you are inserting page 3, *et seq*. The way this guidance/template is formatted should achieve this formatting for you, as long as you copy and paste the template portion (the portion starting on pg. 4 of this document) into a new Word document.

- Use the format "Page # of ##" (e.g., Page 3 of 4).

- Center the page number in the footer.

- Do not enclose in quotes, dashes or parentheses.

o Acronyms: Spell out acronyms or abbreviations in the Subject line, and wherever they are first-used.

o Attachments:

- Each document that accompanies your memo (e.g., a document that will be published in the *Federal Register*, a regulatory impact analysis, an economic analysis) is considered an attachment.

- If attachments are referenced in the body of the text of your memo, type the word "Attachment" or "Attachments (#)" three lines below the body of the memo. For more than one attachment, indicate the number in parentheses.

- If attachments are not identified in the text, type *Attachment* or *Attachments* three lines below the last line of the memorandum body, flush with the left margin. Number and list each attachment on a separate line. If more than one line is needed for any listed attachment, continue the information on a succeeding line aligned with the first character of the name of the attachment.

MEMORANDUM

SUBJECT: [Stage (e.g., Advance Notice of Proposed Rulemaking, Proposed Rule, or Final Rule): Title of Action] (Tier [insert number]; SAN [insert number]; RIN [insert number]) – **ACTION MEMORANDUM**

[Here is an example:
Proposed Rule: CERCLA/EPCRA Administrative Reporting Exemption for Air Releases of Hazardous Substances from Animal Waste (Tier 2; SAN 5117; RIN 2050-AG37) – **ACTION MEMORANDUM**]

FROM: [Insert the name of your Assistant or Regional Administrator]
[Insert "Assistant" or "Regional"] Administrator

THRU: Office of Policy (1806A)
Office of Executive Secretariat (1105A)

TO: [Insert the name of the Administrator]
EPA Administrator (1101A)

[This action memorandum should not exceed four pages, but you may use attachments to provide longer descriptions if necessary. Use plain language throughout. Write this memo so that the Administrator's Office, as well as any future officials who refer to this action's record, can clearly understand the action being taken and its impact. Refrain from copying technical language from your action's preamble or regulatory text.]

PURPOSE
Attached for your signature is a [insert stage (e.g., Advance Notice of Proposed Rulemaking (ANPRM), proposed rule, final rule)]. [In three to five sentences, explain the action and why it was needed. This section should provide some context (i.e., how the rule fits into an overall strategy, agency priority/initiative, or suite of related actions).]

DEADLINE
[Indicate whether any signature or publication deadlines apply. Include this section even if your action does not have a deadline. If there is a deadline, indicate what it is and the type of deadline. It may be a legal deadline (e.g., imposed by a court or by law), an Administration deadline (e.g., identified as a priority action or fulfilling an external commitment), or an internal management deadline (e.g., timed with an event or speech). If no deadline exists, simply state: "No deadlines apply to this action."]

OVERVIEW
[Briefly describe the action, the relevant statute that provides authority for the action and, as appropriate, cover the following points:

- Describe the specific environmental issue(s), public health problem(s) and/or statutory requirements being addressed, and the goal of this action;

- Describe how the regulated community is affected (e.g., performance standards, specific requirements);

- Describe implementation flexibilities, particularly for states and regulated entities;

- Describe key issues, such as any environmental justice concerns or Limited English Proficiency (LEP) concerns related to this action, and how they were addressed;

- Identify other actions underway that will affect this particular program or sector;

- Identify whether the action amends the *Code of Federal Regulations* and, if so, explain what kind of amendment (e.g., procedural); and

- Briefly summarize the history of the action.]

{Many programs elect to use subheadings in this section (e.g., *Authority, Background, Actions Proposed, Key Issues*).}

ANTICIPATED PUBLIC AND STAKEHOLDER RESPONSE

[Describe the type of response anticipated from the various audiences interested or impacted by the action. Identify both the involved stakeholders and the nature of their expected response. Characterize the likely reaction to the action by all interested parties including industry; environmental groups; Congress; state, local and tribal governments; and OMB. Explain what the agency has done to mitigate anticipated adverse reactions.]

INTERNAL DEVELOPMENT AND REVIEW PROCESS

[Identify whether the action was developed under Tier 1, 2 or 3. If the action was Tier 1 or 2, present, in an attachment if preferred, a brief chronology of the development and review process, noting specifically when the workgroup was formed. Note at what stages and for what specific objectives at each stage the workgroup was substantially engaged, including prior to seeking Early Guidance. (The Office of Policy will attach the summary memo from Final Agency Review (FAR).) Describe any noteworthy or innovative collaborative development and review process(es) used internally, and identify those that would be appropriate "best practices" to advance One EPA.]

[Identify program offices or Regions that participated in the development of the action, along with any outstanding issues from the development process and why they cannot be resolved or accommodated. Also, provide the basis for any decision made to not address an identified cross-media impact.]

OMB TRANSACTION

[Identify the determination by the Office of Management and Budget (OMB) (e.g., significant, non-significant, waived) and whether the action went to OMB for review under Executive Order (EO) 12866. If the action went to OMB for review, highlight significant issues resulting from EO 12866 review, including any significant issues raised by other agencies participating in the review. Explain any substantive changes made to the action as a result of recommendations from OMB or the other agencies.]

[If the action is subject to EO 12866 review but OMB waived review (e.g., OMB determined the action was significant but decided not to review it), please indicate whether OMB was otherwise involved with the action (e.g., was briefed) and describe the results of this interaction.]

[If the action is not subject to EO 12866 review, please indicate if OMB was briefed or otherwise involved. Describe the results of this interaction.]

[Note that you will not be able to complete this section until after OMB completes its review of the action; therefore, this section generally will not be complete when you circulate the draft Action Memorandum with the FAR package and the EO 12866 review package to OMB. Do your best to provide what detail you can when circulating the draft memo, however (e.g., it is likely that you can list the OMB determination in this section, even at the draft stage.)]

IMPACTS
[Summarize the costs and benefits of the action – including a discussion of any non-monetized benefits and/or non-quantified benefits – and the results of any economic analyses. As appropriate under individual statutes, explain how cost-benefit analyses helped to shape the approach chosen.]

{Use an attachment to provide any additional economic impact detail and to summarize, as applicable, the regulatory flexibility analysis and Small Business Advocacy Review (SBAR) Panel recommendations. Also, this attachment may describe impacts on affected entities, such as other federal agencies, states, local governments, tribes, paperwork burdens, children's health, environmental justice populations, climate change, etc., that you are likely to discuss in the "Statutory and Executive Order" section of your rule's preamble or in other contexts. Consider using a table to display estimates (i.e., use the Circular A-4 table for economically significant rules.) Reference the attachment in the Impacts section and list the impacts described in the attachment.}

{This attachment should be succinct and focused on salient issues that senior decision-makers in the Office of the Administrator need to know. You may wish to use subheadings in this attachment (e.g., *Environmental Justice, Limited English Proficiency, Small Business Impacts, Federalism Impacts*). Guidance for describing the impacts related to applicable statutes and executive orders can be found in the ADP Library (http://intranet.epa.gov/adplibrary/).}

STAKEHOLDER INVOLVEMENT
[Briefly discuss the role of state, local and tribal government entities and private sector stakeholders (e.g., regulated entities, NGOs, academia) in the development of the action. Summarize the concerns they have raised and what the agency has done to address them, or explain why the agency cannot address them. If applicable, refer to the discussion above or in the attachment on impacts related to EJ concerns, children's health concerns, or other issues described in the Impacts section.]

PEER REVIEW

[If you did not use influential scientific information or a highly influential scientific assessment as defined by the EPA's *Peer Review Handbook* (http://www.epa.gov/peerreview/pdfs/prhandbk.pdf) to support the action, include the following statement in the Action Memorandum: There were no influential or highly influential products supporting this action as defined by the agency's *Peer Review Handbook*.]

[If you did use influential scientific information or a highly influential scientific assessment to support the action, include the following statement: [Insert Name of AAship] has followed the agency's Peer Review Policy with respect to the underlying [influential scientific information or highly influential scientific assessment] supporting this action.]

[You may add any details you think are important, but you generally should not modify this compliance statement. If you used influential scientific information or a highly influential scientific assessment, but were not fully able to meet the Peer Review Policy, explain why.]

RECOMMENDATION

[Recommend an action the Administrator should take (i.e., sign the rule or other document). Here is an example: I recommend that you sign the attached rulemaking.]

[Three lines below the body of the memo, indicate that you have attached a rule for signature, and other documents as appropriate. If your attachment(s) are referenced in the body of the memo, insert either the word "Attachment" or "Attachments (#)," indicating the number of attachments in parentheses. If you have referenced the attachments in the body of your memo, you need not list the names of them here. On the other hand, if attachments are not identified in the body, type "Attachment" or "Attachments" three lines below the last line of the memorandum body, flush with the left margin. Number and list each attachment on a separate line. If more than one line is needed for any listed attachment, continue the information on a succeeding line aligned with the first character of the name of the attachment.]

[NOTE: Where an Action Memorandum accompanies another document (e.g., a rule or *Federal Register* document) to be signed by the Administrator, you should not include a concurrence line at the bottom of the Action Memorandum or anything else that might cause the Action Memorandum to be misinterpreted to be a Decision Memo, which it is not. After signature, the *Federal Register* notice (or other document such as an order) will contain the agency's decision or action (if any). The Action Memorandum is only a pre-decisional briefing document.]

{You may wish to add a "cc" line if you are sending a copy of the action memorandum to others. Do not include a courtesy title such as Mr. or Mrs. The "cc" line should be flush left and two lines below your text or the "Attachment/Attachments" line. Two spaces follow the colon after "cc." If a courtesy copy list is too long to fit in a single column at the bottom of the memorandum, a separate distribution list is permitted and should be referenced in the Attachments.}

APPENDIX E. EXAMPLES OF PEER REVIEW STATEMENTS OF WORK

Note: The examples are provided for reference purposes only. Development of new Statements of Work should reflect current agency policies and procedures, including the "Conflict of Interest Review Process for Contractor-Managed Peer Reviews of EPA HISA and ISI Documents."

Statement of Work: Letter Review

External Peer Review of the EPA's
Markov Chain Nest Productivity Model (MCnest)

STATEMENT OF WORK AND CHARGE TO REVIEWERS

Background

A challenge in the regulation of pesticides is to improve methods for quantifying ecological risk projections in higher-tier risk assessments that can address the "so what" questions about potential changes to wildlife populations. The United States Environmental Protection Agency's (USEPA) Office of Pesticide Programs (OPP) has developed a Terrestrial Investigation Model (TIM) for quantifying the magnitude of acute mortality in birds exposed to a pesticide, but has not adopted a method for quantifying effects to reproductive success. In the current pesticide risk assessment process, results from a pair of laboratory avian reproduction tests are used in calculating risk quotients (RQ) by comparing the reported no-observed-adverse-effect concentration (NOAEC) for the most sensitive measured endpoint(s) with estimates of the maximum dietary exposure expected for a given application rate. As a screening tool, RQs are compared to an established regulatory level-of-concern to categorize the potential for unacceptable risk. Because of the high degree of uncertainty in these simple tools for characterizing risk, RQs typically incorporate conservative or worse-case assumptions about exposure and toxicity to reduce the chances of concluding a chemical has an acceptable level of risk when in fact it does not (i.e., false negative conclusion). Consequently, risk quotients can be used to identify the environmental concentration above which adverse effects to avian reproduction may occur, but they cannot determine the probability or magnitude of potential reproductive effects.

An alternative conceptual framework for interpreting the results of avian reproduction tests was proposed by Bennett et al. (2005). Briefly, it involves linking the types of effects that may occur during each phase of a bird's reproductive cycle (e.g., pair formation, egg laying, incubation, nestling rearing) to selected surrogate endpoints from all three standard avian toxicity tests and relates those effects to the estimated exposure during each phase under a given pesticide-use scenario. Because the great majority of avian reproduction tests do not provide quantitative dose-response information for surrogate endpoints, by necessity the alternative approach is based on a series of phase-specific deterministic decision points – essentially RQs for specific surrogate endpoints at each breeding phase – for determining whether the nest attempt fails or continues. If the estimated exposure during the critical exposure period is less than the established toxicity threshold (e.g., the no-observed-adverse-effect level or NOAEL) for surrogate endpoints at each phase, the nest continues without disruption. However, if exposure exceeds the toxicity threshold for a surrogate endpoint, the nest attempt is assumed to have failed and the female may be able to renest if conditions permit and sufficient time remains in the breeding season. Also, for those species that can produce multiple broods in a single breeding season, females may renest after successful nesting attempts if conditions permit. The simulated performance of a population of females in relation to the timing of pesticide applications is modeled over the course of a full breeding season. Consequently, using this framework, the effects of a pesticide on annual reproductive success are not only a function of the results of avian toxicity tests, but also are quite sensitive to the timing of pesticide applications relative to a species' breeding season and to differences in life history characteristics among species.

A flexible mathematical model, known as the Markov chain nest productivity model or MCnest, has been developed for implementing the conceptual framework of Bennett et al. (2005). It projects estimates of pesticide effects on reproductive success for a broad range of species and can be modified to incorporate either sparse or abundant life-history data. MCnest builds on over 40 years of avian nest-survival modeling in the ornithological literature. This Markov chain model is equivalent to the well-known Mayfield nest-survival model when similar assumptions are. Although the basic version of MCnest was developed to use data from the standardized avian toxicity tests required by OPP, it could be applied to contaminant effects questions in other USEPA Program Offices, though at present the model is not designed to adequately estimate the effects of bioaccumulative chemicals where effects on hatchability and hatchling survival may result from chemical residues accumulated prior to the egg formation period.

Most of the data used in MCnest are in the form of input parameters provided by the model user and represent three categories of input parameters: toxicity threshold values for surrogate endpoints, pesticide application scenarios, and species life history parameters. MCnest uses information for parameterizing toxicity threshold values and application scenarios that is currently available in the risk assessment process. The model user may use default life history parameters from a library of avian species available to MCnest or create new or modified species parameter profiles.

The primary output of MCnest is an estimate of the potential magnitude of pesticide effects to annual reproductive success by calculating the relative difference between scenarios with and without pesticide exposure. It also provides information on which species are at greatest risk under a specific pesticide-use scenario or which application dates have the greatest impact throughout a breeding season. This quantitative estimate of pesticide effects on annual reproductive success is needed for use in population modeling or probabilistic risk assessments.

Scope of Review/Objective Statement

The focus of this review is the MCnest model and its accompanying user's and technical manuals with the objective of providing a written, independent review of the MCnest model and commenting on its ease of use and utility in estimating risk.

The enclosed CD contains a copy of:

- the MCnest model
- an Excel file named 'SpeciesLibrary'
- the MCR Installer
- a User's Manual
- a Technical Manual
- a Species Life History Profiles Manual
- the Bennett et al. (2005) Ecotoxicology publication

The basic version of MCnest focuses on the pesticide risk assessment process of USEPA's Office of Pesticide Programs, though the model could be modified in future versions for application in other USEPA program offices or other regulatory bodies. The document on avian life history profiles is included for background purposes, but is not the focus of this review; once the MCnest model is finalized, the species profiles will be expanded and peer-reviewed separately for use in MCnest. Also, since the purpose of MCnest is to implement the conceptual approach first described by Bennett et al. (2005), that paper is included for background.

Contacts

If you have questions regarding installation or operation of the model, please feel free to contact [Name] at [Phone] or [Email].

If you have questions regarding the review or providing answers to the charge questions, please contact [Name of Peer Review Coordinator] at [Phone] or [Email].

Task Description

1. The contractor shall install the draft MCnest model (including the MCR Installer and the "SpeciesLibrary" excel file) on their computer.

2. The contractor shall perform an independent review of MCnest model and the user's and technical manuals.

3. The contractor shall provide a written evaluation addressing the charge questions, with recommendations, in a report submitted to [the Peer Review Coordinator] no later than [Date].

Government Responsibilities

1. Provide a Statement of Work outlining expectations.

2. Request a conflict of interest statement.

3. Provide a professional services fee (honorarium) if appropriate.

4. Provide a CD including all necessary files.

Milestones/Deliverables and Schedule

The reviewer shall review the MCnest model and the accompanying user's and technical manuals and provide written comments addressing the charge questions to [the Peer Review Coordinator] no later than [Date].

Acceptance Criteria

An independent and unbiased professional review is provided in written form.

Charge Questions

In your written review, please address the following questions. Additional comments and recommendations for improving the model and associated methodology are welcome.

1. The user's manual is intended to introduce all of the currently available features of the MCnest model and allow the model user to start running model simulations. Did you have problems or questions during the installation or operation of the MCnest model? Did you encounter issues that were not explained sufficiently in the user's manual? Do you have any suggestions for improving the user's manual?

2. The technical manual provides background material on the details for how the model operates and how calculations are performed; however, it does not provide guidance on policy-related issues that will need to be addressed for use in a regulatory context. Does the technical manual provide sufficient technical background information for how the model operates and how decisions are made? Are there additional technical issues that should be discussed in the technical manual?

3. The intent of the basic version of MCnest is to implement a breeding phase-specific approach for quantifying pesticide effects on avian reproductive success that is general enough to be applied to a broad range of species life history strategies. Does the technical manual adequately explain the selection and use of surrogate endpoints? Do the manuals

adequately explain how choices made for input parameters might affect the model results?

4. Despite the limitations of both toxicity test data and life history information, does the model provide a basis for quantifying the magnitude of change to reproductive success from pesticide exposure that adds value beyond the current use of risk quotients? Does the model adequately implement the breeding phase-specific approach for quantifying pesticide effects on reproductive success?

5. Beyond the basic model outputs provided, are there additional outputs (e.g., graphs, data summaries) that would be useful for understanding the simulation results or interpreting differences among simulations?

6. Work is underway to include more detailed exposure estimation and improved methods for defining the length of a breeding season in future versions of MCnest. Are there additional features or issues that you believe should be addressed in MCnest?

Reporting Requirements

Please provide your written comments to [the Peer Review Coordinator] by [Date], answering the questions specified above. The review may be sent by regular mail to the address below, by email to [Email] or by FAX to [Number].

We sincerely thank you for your input to our peer review process.

Peer Review Coordinator
Address
Phone
Email

Statement of Work: Contractor-Managed Peer Review

STATEMENT OF WORK
U.S. EPA Environmental Economics Peer Review

BACKGROUND INFORMATION:

The National Center for Environmental Economics (NCEE) is located in the Office of Policy, Economics and Innovation and serves as a center of expertise for cutting-edge research and analysis in environmental economics. NCEE's primary function is to assist the Environmental Protection Agency's (EPA's) program and staff offices in applying sound economic science in the development of analyses that support the Agency's actions. NCEE conducts and supervises a wide array of research and development on economic analytic methods, and provides guidance and support for performing economic analyses throughout the Agency. NCEE serves as an information resource for EPA, other government departments and agencies, and the public on benefit-cost analyses, economic impact models and measurement, and economic incentive measures.

Peer review is an important component of the scientific process. It provides a focused, objective evaluation of work products, and the criticism, suggestions and new ideas provided by the peer reviewers stimulate creative thought, strengthens the reviewed document and confer credibility on the product. Comprehensive, objective peer reviews leads to good science and product acceptance within the scientific community.

PURPOSE:

NCEE (and economists throughout the Agency) routinely create work products that require peer review. The purpose of this contract is to procure peer review services from a contractor that is able to perform peer review of a variety of environmental economic work products. The economic work products for peer review required under this contract, as described in EPA's *Peer Review Handbook*, 3rd Edition (http://www.epa.gov/peerreview/pdfs/peer_review_handbook_2006.pdf) are as follows:

1. Economic and financial methodologies that will serve as a principal method or protocol used to conduct economic analyses within a program;

2. Unique or novel applications of existing economic and financial methodologies , particularly those that are recognized to be outside of mainstream economic practices;

3. Stated preference (e.g., contingent valuation) and revealed preference surveys (e.g., recreational travel cost surveys) developed to assist in the economic analysis of a regulation or program ;

4. National surveys of costs and expenditures for environmental protection (e.g., financial needs surveys, pollution abatement expenditures surveys);

5. Meta-analyses (i.e., re-analyses of existing published literature and supporting data on the measurement of economic benefits, costs and impacts);

6. Data and analytical models underlying economic analyses, particularly those

supporting economically significant rules, if the models and corresponding use of the data have not been previously subject to adequate peer review; and

7. Applications for research grants.

Note that the above list omits two (2) kinds of economic work products described in the Peer Review Handbook that are typically peer reviewed: internal Agency guidance for conducting economic and financial analysis; and broad-scale economic assessments of regulatory programs, such as those required by Congressional mandates (e.g., the Clean Air Act reports to Congress on benefits and costs). These major work products would usually be reviewed by EPA's Science Advisory Board Environmental Economics Advisory Committee or an equivalent body, and are not in the scope of this Statement of Work (SOW).

Examples of current and/or previous work products produced by NCEE in need of peer review are as follows:

1. NCEE recently published the Handbook on the Benefits, Costs, and Impacts of Land Cleanup and Reuse (http://yosemite.epa.gov/ee/epa/eed.nsf/pages/LandHandbook.html), which is 126 pages and which was peer reviewed by a panel of seven (7) environmental economists.

2. NCEE is also leading a large effort to value the benefits of the Chesapeake Bay Total Maximum Daily Load, including a stated preference survey, a hedonic analysis, commercial and recreational fishing benefits, protecting drinking water in groundwater wells, other ancillary benefits, and a benefit transfer exercise. NCEE expects that each component of this analysis will result in a report of approximately 50 pages that would need to be peer reviewed by three (3) outside economists.

3. NCEE routinely issues proposals for research grant applications; these proposals are usually 10-15 pages in length with some supporting information. All research grant proposals must be externally peer-reviewed and NCEE prefers three (3) external reviewers per proposal.

The purpose of this contract is to provide peer review for the whole variety of environmental economics work products from NCEE and EPA. Several offices within EPA will utilize this contract, for in-scope work, based upon NCEE projections and future requirements.

TECHNICAL SUPPORT REQUIREMENTS:

The contractor shall perform the following tasks in support of this contract:

A. Peer Review Services

The contractor shall perform scientific/technical peer reviews of documents and materials related to the full breadth of Agency work products and grant/cooperative agreement proposals pertaining to environmental economics. The peer reviews may occur by mail or email; via telephone or video conferences; or during in-person meetings.

When conducting peer reviews, the contractor shall follow EPA's *Peer Review Handbook*, 3rd edition (EPA 1OO/B-06/002, January 2006, which is provided at the following website:

http://www.epa.gov/peerreview/pdfs/peer_review_handbook_2006.pdf; and *Addendum* (2009) at the following link: http://www.epa.gov/peerreview/pdfs/spc_peer_rvw_handbook_addendum.pdf or the most recent rendition of that Handbook, to the extent that the subject of the review is covered by EPA's *Peer Review Handbook*, 3rd Edition and the *Addendum*.

1. Identify and Recruit Qualified Reviewers

The number of reviewers required and their qualifications will be determined during contract performance and will be provided by the Contracting Officer's Representative (COR); the reviewer(s) qualifications may vary depending upon the technical nature of the work product. The minimum qualifications for a peer reviewer of the products encompassed in this contract are a Ph.D. in economics, environmental economics, agricultural economics, or a related field. Interdisciplinary projects may, in some cases, require expertise (as demonstrated by a Ph.D.) in a different field; any such cases will be indicated by the COR.

The appropriate expertise, knowledge, and experience necessary for individual peer reviewers will be indicated by the COR according to the following:

> Level 1 reviewers will have engaged in relevant research as evidenced by at least one peer-reviewed journal publication in the subject of the review.

> Level 2 reviewers will have engaged in relevant research as evidenced by at least three peer-reviewed journal publications in the subject of the review; or by at least one peer-reviewed journal publication in the subject of the review and by serving as the principal investigator for a research project comparable to the product being reviewed.

> Level 3 reviewers will have engaged in relevant research and achieved standing in the field as evidenced by at least four peer-reviewed journal publications in the subject of the review; by serving as the principal investigator for at least one research project comparable to the product being reviewed; and by achieving recognition in the field as reflected by awards, and other honors received from scientific and professional organizations (e.g., an AERE or AAAS Fellow), distinguished or named professorships, journal editorships, or appointment to high-level review committees (such as the National Research Council or Science Advisory Board).

Prior to the performance of a peer review, the contractor shall submit to the COR and the CO a Statement of Conflict of Interest for each reviewer in addition to a complete list of all prospective reviewers within two (2) weeks of the contractors receipt of the number of reviewers and their qualifications from the COR. For each prospective reviewer, the submission shall include: (1) a short academic and professional biography, and a brief paragraph concerning the reviewer's technical expertise in support of the reviewer's selection; and (2) information concerning the reviewer's availability and willingness to provide the review within the specified time frame. The CO will verify that the list of reviewers conforms to the number and qualifications of reviewers provided to the contractor and required for the review. The contractor will be notified by the CO with a determination of consent with regard to the list of proposed reviewers. NCEE will not be involved in the selection of individual peer reviewers.

Within three (3) working days of receiving the CO's consent that the contractor's proposed list of reviewers conforms to the peer review's specifications and Statement of Conflict of Interest, the contractor shall select and enlist the services of reviewers.

It is the responsibility of the contractor to ensure that all peer reviews are conducted in a manner to avoid all actual or potential, substantial conflicts of interest, or the appearance of substantial conflicts to the maximum extent possible. Prior to conducting a peer review, the contractor shall ensure that each reviewer is free of any actual or potential conflict of interest (COI), or the appearance of any substantial conflict that are direct and substantial enough as to rule out a particular reviewer. Any particular COI, or appearance of loss of impartiality (see Chapter 5), must be disclosed by the contractor with a description of the actions the contractor has taken, or proposes to take, to avoid, mitigate, or neutralize the COI or appearance of loss of impartiality. Assurance of impartiality of each reviewer must be provided by the contractor to EPA.

2. Submission of Written Comments

Each review will be directed by a charge (including general and specific questions, evaluation criteria, or similar instructions to peer reviewers) that will be provided by the COR. (See Section 3.2 of the Peer Review Handbook for a description of charges.)

After completing the review, the contractor shall submit the peer review panel's written comments in final form, along with all supporting materials, such as additional references or suggested approaches, to the appropriate EPA personnel. Review packages submitted by the contractor to EPA shall include: (1) written general comments; (2) specific changes or revisions required to improve clarity; (3) scientific changes or revisions required to improve the clarity and/or the scientific accuracy of the documents or products; (4) any new data that might contribute to the derivation of improved processes and procedures; (5) other scientific and technical materials that may be pertinent; and (6) any other materials necessary to complete the peer review record. The contractor shall also be readily available to clarify any peer review comments and recommendations the EPA poses within a week of submission of the review package.

The contractor shall submit these documents to the COR in final form. All final peer reviews submitted shall include copies of the literature cited or make reference to the citations in the document for the COR to verify and approve.

3. Submission of Panel Recommendations

For reports that include peer review panel recommendations, the contractor shall: (1) explain and rank the policy or action alternatives; (2) describe the procedures used to arrive at the recommendations; (3) summarize the substance of the peer review panel's deliberations; (4) summarize any peer review panel dissenting views; (5) list the sources relied upon; and (6) provide any other information necessary clarify the methods and considerations upon which the recommendations are based.

The contractor shall submit these documents to the COR in final form. All final peer reviews submitted shall include copies of the literature cited or make reference to the citations in the document for the COR to verify and approve.

B. Workshop and Meeting Support Requirements
The contractor shall perform the following activities relative to peer review meetings, meetings that are not explicitly peer review, as well as for scientific workshop/workshop support, in support of this contract:

1. Pre-Workshop/Meeting Support

The contractor shall:

a) Identify attendees. Organize and provide support in arranging workshops, meetings, and presentations by individuals to address issues and concerns raised by the peer review and/or as requested;

b) Arrange for workshops and meetings to be held at EPA office locations or other geographical sites, as specified during contract performance;

c) Arrange for facilities necessary to support required equipment, agenda development, and other logistical support , including: tape recording, audiovisual, computer, photo-copying, and operation of audiovisual equipment, microphones, and lighted pointers (all photocopying shall adhere to the clause Printing, (EPAAR 1552.208-70) (DEC 2005));

d) Select hotel and arrange for rooms for workshop participants, as necessary;

e) Develop the registration process and the materials needed for pre- workshop/meeting and onsite activities, e.g., registration and distribution of workshop/meeting materials, agendas, literature, information pamphlets, etc. to participants;

f) Inspect meeting and workshop site with site personnel, checking all facilities, furniture, equipment, and signs to ensure facilities are appropriate and sufficient to handle meeting/workshop and attendees' requirements; and

g) Provide identification badges for workshop and meeting attendees.

h) The contractor shall clearly identify itself as an EPA contractor. When in attendance at meetings, contractor personnel shall wear identification that is different than the badges used by seminar attendees or Agency personnel attending or speaking at the meeting. Contractor personnel shall identify themselves as such when placing calls in conjunction with the SOW.

i) Arrange teleconferences for planning purposes, peer review panels, or similar purposes.

2. Post-Workshop/Meeting Support

The contractor shall:

a) Obtain all post-meeting/workshop comments, collect and compile all comments and suggested document revisions, transcribe meeting proceedings where required , and obtain all new hard copy references with distribution of copies as specified during contract performance; and

b) Distribute draft proceedings summaries to the peer reviewers and/or participants for comment. Distribute revised proceedings summaries to the COR for review to ensure completeness and clarity before development of final document(s).

DELIVERABLES AND SERVICES

1. The contractor shall deliver complete comments, as specified during contract

performance, to assure rapid assimilation and timely action by EPA. Deliverables will be used to improve the quality of planned and current research projects and to assess the scientific and technical accuracy of completed and current work before dissemination outside EPA. Specific deliverables will be specified during contract performance. In providing deliverables , the contractor shall: provide high quality peer reviews and workshops in the research areas specified during contract performance;

2. Maintain the capability to provide such peer reviews and workshops as needed , document qualifications of personnel, and ensure performance of the work in accordance with EPA guidance;

3. Disseminate existing EPA supplied and specified documents, as referenced during contract performance;

4. Arrive at firm conclusions and/or recommendations, and provide supporting documentation and/or analyses to EPA;

5. Coordinate peer review findings with EPA and other selected individuals through teleconferences, workshops , and/or meetings involving the COR and any other specified EPA personnel, to clarify specific scientific points made by a peer review panel , and to document views and scientific judgments made by peer reviewers;

6. Provide a full and accurate accounting of all work ordered, as required;

7. Document the procedures used to ensure that all specifications required for a given review are met;

8. Maintain a record of ongoing and completed peer reviews, and devise a system for documenting all peer reviews conducted; an electronic copy of a progress report shall be sent to the COR and the CO.

9. Provide follow-up information to peer reviewers, to the COR and the CO;

10. Certify that, to the best of the contractor's knowledge and belief, no actual or potential conflicts of interest, or appearance of substantial conflicts exist, in accordance with contract requirements; and

11. Provide pre-meeting, meeting, and post-meeting workshop support, including planning, arranging, administering, and conducting required workshops and/or meetings.

Copies of all deliverables shall be sent to the COR in an electronic format (i.e., MS Word or MS Excel, MS Office 97 or a later version), along with a portable document file (.pdf) copy. The deliverable must include, but not be limited to, the peer review or workshop title. In all matters, the contractor shall perform in a manner that will ensure consistency of procedure and practice in support of the requirements of this SOW, and shall ensure consistent completion of all deliverables in accordance with the contract.

APPENDIX F. GUIDANCE ON REQUESTING A REVIEW
BY THE SCIENCE ADVISORY BOARD

Introduction

Each year, the Deputy Administrator invites the EPA's senior leadership to identify requests for advice and peer review from the agency's independent Science Advisory Board (SAB), Clean Air Scientific Advisory Committee (CASAC), and other advisory committees. Significant scientific and technical issues related to the Administrator's priorities are topics most appropriate for consideration by these science advisory committees, which are supported by the SAB Staff Office in the Office of the Administrator. This appendix provides guidance for identifying and nominating requests for SAB and CASAC review. More detailed information about the functions and advisory process for the committees is available at http://www.epa.gov/sab.

Background

The SAB and CASAC provide mechanisms for the EPA to receive external peer review and other advice designed to make a positive difference in producing and using science at the agency.

The SAB has a broad congressional mandate to provide independent advice and peer review to the EPA Administrator on the scientific and technical aspects of environmental issues. Section (c)(1) of the Environmental Research, Development, and Demonstration Authorization Act ([ERDDAA], 42 U.S.C. § 4365] states that "at the time any proposed criteria document, standard, limitation, or regulation … under any … authority of the Administrator, is provided to any other Federal agency for formal review and comment, shall make available to the Board such proposed criteria document, standard, limitation, or regulation, together with relevant scientific and technical information in the possession of the Environmental Protection Agency on which the proposed action is based."

The CASAC provides independent advice to the EPA Administrator on the technical bases for the EPA's national ambient air quality standards program, including peer review of Integrated Science Assessments, Risk and Exposure Assessments, and Policy Assessments for criteria air pollutants.

These advisory committees generally provide advice on high-priority scientific and technical issues in written form, either as peer reviews of final draft technical reports (e.g., guidelines, assessments, research strategies) or work products (e.g., analytical methods, models, databases) or advisories (written advice on works in progress). In some cases, where the EPA is committed to interact with the SAB iteratively in developing a scientific product or activity, advisory members may provide an initial consultation to provide advice at an early stage in a science activity. Such a consultation will be followed at a later stage by an advisory or peer review report. The SAB also may provide oral rapid consultative advice in the event of an emergency, such as a natural disaster, and may conduct *de novo* studies on emerging science issues or overarching topics of importance to the EPA.

Because resources are always limiting, the SAB Staff Office uses several criteria for selecting project proposals proposed by the agency. Advisory project proposals best suited for consideration by the SAB and CASAC are those that meet several of the following criteria:

- General Criterion:

  o Provides an opportunity to make a difference in the science that supports the agency's mission.

- Client-Related Criteria:

  o Supports major regulatory or risk management initiatives.

  o Serves leadership interests (e.g., the Administrator, Congress).

  o Supports EPA strategic priorities.

- Science-Driven Criteria:

  o Involves scientific approaches that are new to the EPA.

  o Addresses areas of substantial uncertainties.

- Problem-Driven Criteria:

  o Involves major environmental risks.

  o Relates to emerging environmental issues.

  o Exhibits a long-term outlook.

- Organizational Criteria:

  o Serves as a model for future agency methods.

  o Requires the commitment of substantial resources to scientific or technological development.

  o Transcends organizational boundaries, within or outside the EPA (includes international boundaries).

  o Strengthens the agency's basic capabilities.

In addition, the SAB Staff Office considers the overall mix of the nominated project proposals for a specific fiscal year, as well as the time and available resources needed to take on the projects.

Process for Submitting Nominations

Any office desiring to take a product, activity or issue to the SAB for a peer review, advisory project nomination or consultation is requested to complete the two-step process described below.

Step 1 – Project Identification and Nomination. Each year, the Assistant Administrators and Regional Administrators are asked to send the SAB Staff Office Director a memorandum that lists all advisory project nominations, with the highest priority nominations for the next fiscal year identified

Step 2 – Electronic Project Sheet. Nominators are asked to submit an electronic project sheet for each individual project to be considered for SAB or CASAC attention. The project sheets are created after establishing or updating the related Peer Review Project or Science Activity in the agency's Science Inventory. The Science Inventory entries must be approved by their Peer Review Coordinator or Science Activity Coordinator for the specific program or regional office. Project sheets should be filled out for all desired projects, including previously submitted projects for which no project planning meeting has occurred between the program or regional office and the SAB Staff Office Director. The electronic Project sheet may be accessed and completed through the SAB Product Database; contact the SAB Staff Office for information about how to access the database to create a project sheet. The information fields required for the electronic project sheet are provided in Table E-1.

Process for Keeping Informed About the Decisions Made

After receiving project nominations, the SAB Staff Office will discuss project priorities with each EPA program and regional office. The Staff Office also discusses project priorities with the chartered SAB, which includes the chair of the CASAC. The SAB Staff Office will consult with the EPA Administrator, Deputy Administrator and Science Advisor to the Administrator to develop an annual operating plan that includes the highest priority projects. Additionally, projects may be added or deleted at any time during the fiscal year, as requested by EPA senior management, Congress and the SAB.

The SAB Staff Office will identify a point of contact for each advisory request and keep requesting offices informed about the status of advisory activities. The SAB website (www.epa.gov/sab) provides current information about advisory activities once they are announced to the public. This information includes the *Federal Register* notice announcing the advisory activity, information about panel formation, public meetings, draft reports, quality review by the chartered SAB (which reviews and approves all advisory reports), final reports to the Administrator, and the Administrator's responses to final reports.

The SAB and CASAC are federal advisory committees subject to the Federal Advisory Committee Act. For more information about how these advisory committees operate and the roles of the public and the agency in that process, please see *Advisory Committee Meetings and Report Development: Process for Public Involvement* (EPA-SABSO-04-001).

EPA Staff with questions about the SAB Product Database or the process for submitting nominations to the EPA SAB may contact the SAB Staff Office.

Table F-1. Information Fields for Science Advisory Board (SAB) Project Sheet

1. Project Title
2. Project Short Title
3. Fiscal Year SAB Activity Desired to Begin
4. Quarter SAB Activity Desired to Begin
5. Requesting Assistant Administrator/Regional Administrator
6. Requesting Office
7. Requesting Official (Division Director or above)
8. Requesting Official's Title
9. Program Contact
10. Program Contact's Phone
11. Program Contact's Mail Code
12. Background for This Advisory Activity
13. Tentative Charge
14. Applicable GPRA Goal and Objective
15. Description of and Citation for Any Legal Obligation/Directive for SAB Review:
16. Principal Interested and Affected Parties
17. Type of SAB Advice Requested
18. Why Should the SAB Advise on This Project?
19. Disciplinary Expertise
20. Budget:
    - FY
    - Extramural Budget
    - FTE
21. Past Peer Reviews
22. Quality Management/Quality Assurance:

APPENDIX G. EPA FEDERAL ADVISORY COMMITTEES THAT PERFORM SCIENTIFIC PEER REVIEW

The Federal Advisory Committee Act (FACA; 5 U.S.C., App. 2) is a statute designed to ensure that the Congress and the public are kept informed of the activities of advisory committees that report to the executive branch of the federal government. Key provisions of the law are that committees must have balanced membership in terms of points of view for the tasks to be performed, meetings are to be announced ahead of time and open to interested members of the public, detailed meeting minutes are to be kept, and all materials presented to or prepared by or for the committees are to be made available to the public. In addition, all federal advisory committees must have a formal charter filed with the head of the agency and the Congress. For more information on Federal Advisory Committees at the EPA, see http://www.epa.gov/ocem/faca/.

The following scientific advisory committees have been established at the EPA to provide scientific advice and peer review:

- **EPA Science Advisory Board (SAB)**: a statutory committee established under the Environmental Research, Development, and Demonstration Authorization Act (ERDDAA, codified at 42 U.S.C. § 4365) to provide independent advice and peer review to the EPA's Administrator on the scientific and technical aspects of environmental issues, including the adequacy and scientific basis of any EPA proposed criteria document, standard, limitation or regulation. The SAB reports directly to the EPA Administrator. For more information on the SAB, see http://www.epa.gov/sab.

- **EPA Clean Air Scientific Advisory Committee (CASAC)**: a statutory committee established under the Clean Air Act (42 U.S.C. § 7409(c)(2)) to provide independent advice on the scientific and technical aspects of air quality criteria and standards, research related to air quality, sources of air pollution, and strategies to attain and maintain air quality standards and prevent significant deterioration of air quality. The CASAC reports directly to the EPA Administrator. For more information on the CASAC, see http://www.epa.gov/casac.

- **FIFRA Scientific Advisory Panel (SAP)**: a statutory committee established under the Federal Insecticide, Fungicide, and Rodenticide Act (FIFRA, 7 U.S.C. § 136w) that provides advice, evaluations and recommendations on pesticides and pesticide-related issues relating to the impact on health and the environment of the EPA's pesticide-related regulatory actions. The FIFRA SAP reports to the EPA Administrator through EPA's Assistant Administrator for the Office of Chemical Safety and Pollution Prevention (OCSPP). For more information on the FIFRA SAP, see http://www.epa.gov/scipoly/sap.

- **EPA Board of Scientific Counselors (BOSC)**: a discretionary committee established by the EPA to provide advice and recommendations on technical and management issues relating to the Office of Research and Development's (ORD) research program. As appropriate, the BOSC coordinates its work with the SAB. The BOSC reports to the EPA Administrator through the Assistant Administrator for the Office of Research and Development, in consultation with the Administrator's Science Advisor. For more information on the BOSC, see http://www.epa.gov/OSP/bosc.

- **Human Studies Review Board (HSRB):** a statutory committee that provides scientific or policy advice to the EPA on the scientific and ethical aspects of human subjects research. The HSRB reports to the EPA Administrator through the EPA's Science Advisor. For more information about the HSRB, see http://www.epa.gov/osa/hsrb.

APPENDIX H. EXAMPLES OF PEER REVIEW CHARGES

It should be noted that certain questions posed in charges can be responded to with a yes or no answer. Clearly, this is not the type of response the agency generally wants; therefore, it is important to phrase charge questions carefully to ensure a fully satisfactory and thoughtful response. Where a yes or no answer might be expected, charge questions should ask for a full explanation supporting the yes or no answer.

Charges can run the gamut from rather simplistic to highly complex, depending on the nature of the review. The examples shown here cover a variety of types. Examples 1 through 3 have less complex questions and are looking for the overall quality of the efforts. Examples 4 and 5 have numerous technical questions that need to be addressed and are, therefore, more complex in their nature.

Other charges that have been used can be found on the Science Advisory Board (SAB) website at http://www.epa.gov/sab and the Scientific Advisory Panel (SAP) website at http://www.epa.gov/scipoly/sap/index.htm.

Charge Example 1: HISA Example

US ENVIRONMENTAL PROTECTION AGENCY
OFFICE OF RESEARCH AND DEVELOPMENT
National Coastal Condition Report IV

Charge to the Peer Reviewers

This document represents a collaborative effort among EPA's Office of Water and Office of Research and Development, NOAA, and US Fish and Wildlife Service. Our objective is to provide regional and national assessments of the condition of coastal waters of the U.S.

Background:

The National Coastal Condition Reports represent collaboration among EPA (OW and ORD), NOAA, USFWS, and coastal state agencies. The first National Coastal Condition Report (NCCR I), published in 2001, reported that the nation's coastal resources were in fair condition. The NCCR I used available data from 1990 to 1996 to characterize approximately 70% of the nation's coastal resources. The second National Coastal Condition Report (NCCR II) was based on data from 1997 to 2000 representative of 100% of coastal area in the contiguous 48 states and Puerto Rico, and showed that the nation's coastal waters continued to be in fair condition. The 3rd National Coastal Condition Report (NCCR III) assessed condition of the nation's coastal waters, including Alaska and Hawaii, based primarily on NCA data collected in 2001 and 2002, and indicated that the condition remained fair. For the first time, NCCR III also included comparison of changes in condition from 1990 to 2002, presented for the nation's coastal waters and by region.

The National Coastal Condition Report IV (NCCR IV) is the fourth in a series of environmental assessments of U.S. coastal waters and the Great Lakes. The report includes assessments of all the nation's coastal waters in the contiguous 48 states and Puerto Rico, south-eastern Alaska, Hawaii, the U.S. Virgin Islands, Guam, and American Samoa. The NCCR IV presents four main types of data: (1) coastal monitoring data, (2) coastal ocean/offshore monitoring data, (3) offshore fisheries data, and (4) beach assessment and fish advisory data. The NCCR IV relies heavily on coastal monitoring data from EPA's National Coastal Assessment (NCA) to assess coastal condition by evaluating five indices of condition—water quality, sediment quality, benthic community condition, coastal habitat loss, and fish tissue contaminants. Coastal waters are valuable from both an environmental and economic perspective. These waters are vulnerable to pollution from diverse sources. EPA expects that this report on the condition of coastal waters will support more informed decisions concerning protection of this resource and will increase public awareness about the extent and seriousness of pollution in these waters.

The overall condition of the nation's coastal waters is fair, using five key indices of ecological health [water quality index (including dissolved oxygen, chlorophyll a, nitrogen, phosphorus, and water clarity), sediment quality index (including sediment toxicity, sediment contaminants, and sediment total organic carbon), benthic index, coastal habitat index, and a fish tissue contaminants index]. For each of these five key indices, a score of good, fair, or poor was assigned to each coastal region of the U.S. These ratings were then averaged to create overall regional and national scores illustrated using "traffic light" color scoring.

Purpose:

The purpose of this review is to obtain expert feedback and comments on the draft "*National Coastal Condition Report IV*." In your review, please provide written responses to the questions below. Additional comments and recommendations for improving the report and associated methodology are also welcome.

Charge Questions:

1) Are the methods used to assess coastal condition supported by sound scientific principles?

2) Selection and use of coastal monitoring indicators are described in Chapter 1 (the Introduction). Do the coastal monitoring indicators used to assess coastal condition nationally and regionally and do the criteria for ranking condition as good, fair, or poor reflect the primary environmental concerns of state, regional, and national resource managers?

3) Are the report's conclusions supported by the analyses and results?

4) Are the conclusions regarding changes in coastal condition over time supported by the data and analyses presented?

5) Does this report represent an important contribution to the state of the science for assessment of coastal waters?

6) Do the four approaches to assessing coastal condition (i.e., coastal monitoring data, coastal ocean/ offshore monitoring data, offshore fisheries, and assessment and advisory data) clearly represent aspects of coastal condition that are informative and not redundant?

7) Are the shortcomings of available data and assessment approaches clearly articulated?

8) This report is quite lengthy. For those reviewing the entire document or individual chapters, do you have any recommendations for omitting parts of this report to shorten the length?

9) Please discuss any controversies that may be raised by the conclusions presented in this report.

Please provide written comments to EPA's Peer Review Coordinator, [name], by [date]. Your review may be sent by regular mail to the address below, by e-mail to [email] or by fax to [fax number].

If you have any questions concerning the draft report or the charge, please contact me at [phone number] or [email]. We sincerely thank you for your input to this important peer review.

Charge Example 2: Charge for a Letter Review for an Economic Analysis

Letter Reviews of Chapter 10 of the EPA Economic Analysis Guidelines

Document: Chapter 10 entitled "Environmental Justice, Children's Environmental Health, and Other Distributional Considerations" -- approximately 37 pages total

Task: A letter peer review of this report by three (3) external economists with Level 3 expertise.

- At least one (1) economist will have specific expertise with environmental justice analysis.

- At least one (1) economist will have specific expertise with public health and distributional analysis.

- At least one (1) economist will have familiarity with environmental regulatory impact analysis and how risk assessment information is used to inform regulatory impact analysis.

Deliverable: Written review comments from each reviewer

Deadline: 4 weeks from receipt of document for review

Charge Questions for External Peer Review of Chapter 10: Environmental Justice, Children's Environmental Health, and Other Distributional Considerations

1. Please provide your overall impressions of the clarity and technical accuracy of the discussion in the chapter for analyzing and presenting quantitative information about the distributional effects of environmental regulations with regard to race and income.

2. A brief overview of the environmental justice literature from the economics field is provided. Are there any pertinent citations that should be added to the discussion?

3. The chapter presents a suite of methods to describe the distributional effects of environmental regulations. Please comment on the technical accuracy and clarity with which each method is described. Are there any methods or relevant literature that should be added to the discussion or deleted?

4. The chapter primarily describes the use of Census data for conducting analyses. Are there additional data sources that should be included?

5. Does the text box on Social Welfare Functions and Inequality Indices provide a reasonable discussion of the available literature and challenges in using these indices in the context of measuring changes in the distribution of environmental quality? Are the conclusions regarding the use of SWFs and inequality indices in this context technically accurate and scientifically grounded?

6. The chapter recommends that all economically significant rules include summary statistics on EJ, with supplemental methods for measuring and estimating EJ impacts as appropriate for the action. Please comment on whether this is a reasonable approach for presenting the analytic

results, and if there are other recommended ways to present the analytic results to inform decision-making?

7. Please provide your overall impressions of the clarity and technical accuracy of the discussion of elderly, children and intergenerational equity. Should other methods or considerations be added to this discussion?

8. Are there additional equity dimensions that should be considered in this chapter?

Charge Example 3: ISI (Panel Review)

Charge to External Reviewers for the IRIS Toxicological Review of Biphenyl

September 2011

Introduction

The U.S. Environmental Protection Agency (EPA) is seeking an external peer review of the draft Toxicological Review of Biphenyl that will appear on the Agency's online database, the Integrated Risk Information System (IRIS). IRIS is prepared and maintained by the EPA's National Center for Environmental Assessment (NCEA) within the Office of Research and Development (ORD). The existing IRIS assessment for biphenyl includes a chronic reference dose (RfD) posted in 1989 and a cancer weight-of-evidence descriptor posted in 1991. The external review draft Toxicological Review of Biphenyl includes an RfD and a cancer assessment.

Charge Questions

Below is a set of charge questions that address scientific issues in the draft Toxicological Review of Biphenyl. Please provide detailed explanations for responses to the charge questions. EPA will also consider reviewer comments on other major scientific issues specific to the hazard identification and dose-response assessment of biphenyl. Please identify and provide the rationale for approaches to resolve the issues where possible. Please consider the accuracy, objectivity, and transparency of EPA's analyses and conclusions in your review.

General Charge Questions:

1. Is the Toxicological Review logical, clear and concise? Has EPA clearly presented and synthesized the scientific evidence for noncancer and cancer health effects of biphenyl?

2. Please identify any additional peer-reviewed studies from the primary literature that should be considered in the assessment of noncancer and cancer health effects of biphenyl.

Chemical-Specific Charge Questions:

(A) Oral reference dose (RfD) for biphenyl

1. A developmental toxicity study of biphenyl in Wistar rats (Khera et al., 1979) was selected as the basis for the derivation of the RfD. Please comment on whether the selection of this study is scientifically supported and clearly described. If a different study is recommended as the basis for the RfD, please identify this study and provide scientific support for this choice.

2. A developmental effect in Wistar rats (i.e., fetal skeletal anomalies) was concluded by EPA to be an adverse effect and was selected as the critical effect for the derivation of the RfD. Please comment on whether the selection of this critical effect and its characterization is scientifically supported and clearly described. If a different endpoint is recommended as the critical effect for deriving the RfD, please identify this effect and provide scientific support for this choice.

3. Benchmark dose (BMD) modeling was conducted using the incidence of litters with fetal skeletal anomalies to estimate the point of departure (POD) for derivation of the RfD. Has the modeling been appropriately conducted and clearly described based on EPA's draft *Benchmark Dose Technical Guidance Document* (U.S. EPA, 2000)? Is the choice of the benchmark response (BMR) for use in

deriving the POD (i.e., a BMR of 10% extra risk of the incidence of litters with any fetal skeletal anomalies) supported and clearly described?

4. Please comment on the rationale for the selection of the uncertainty factors (UFs) applied to the POD for the derivation of the RfD. Are the UFs appropriate based on the recommendations described in *A Review of the Reference Dose and Reference Concentration Processes* (U.S. EPA, 2002; Section 4.4.5) and clearly described? If changes to the selected UFs are proposed, please identify and provide scientific support for the proposed changes.

(B) Inhalation reference concentration (RfC) for biphenyl

1. The draft Toxicological Review of Biphenyl did not derive an RfC. Has the justification for not deriving an RfC been clearly described in the document? Are there available data to support the derivation of an RfC for biphenyl? If so, please identify these data.

(C) Carcinogenicity of biphenyl

1. Under EPA's *Guidelines for Carcinogen Risk Assessment* (U.S. EPA, 2005; www.epa.gov/iris/backgrd.html), the draft Toxicological Review of Biphenyl concludes that the database for biphenyl provides "suggestive evidence of carcinogenic potential" by all routes of exposure. Please comment on whether this characterization of the human cancer potential of biphenyl is scientifically supported and clearly described.

2. EPA has concluded that biphenyl-induced urinary bladder tumors in male rats is a high-dose phenomenon involving sustained occurrence of calculi in the urinary bladder leading to transitional cell damage, sustained regenerative cell proliferation, and eventual promotion of spontaneously initiated tumor cells in the urinary bladder epithelium. Please comment on whether this determination is scientifically supported and clearly described. Please comment on data available that may support an alternative mode of action for biphenyl-induced urinary bladder tumors.

3. EPA has concluded that there is insufficient information to identify the mode(s) of carcinogenic action for biphenyl-induced liver tumors in mice. Please comment on whether this determination is appropriate and clearly described. If it is judged that a mode of action can be established for biphenyl-induced mouse liver tumors, please identify the mode of action and its scientific support (i.e., studies that support the key events, and specific data available to inform the shape of the exposure-response curve at low doses). *Oral Slope Factor (OSF)*

4. A two-year cancer bioassay of biphenyl in BDF1 mice (Umeda et al., 2005) was selected as the basis for the derivation of the OSF. Please comment on whether the selection of this study is scientifically supported and clearly described. If a different study is recommended as the basis for the OSF, please identify this study and provide scientific support for this choice.

5. The incidence of liver tumors (i.e., adenomas or carcinomas) in female mice was selected to serve as the basis for the derivation of the OSF. Please comment on whether this selection is scientifically supported and clearly described. If a different cancer endpoint is recommended for deriving the OSF, please identify this endpoint and provide scientific support for this choice.

6. Benchmark dose (BMD) modeling was conducted using the incidence of liver tumors in female mice in conjunction with dosimetric adjustments for calculating the human equivalent dose (HED) to estimate the point of departure (POD). A linear low-dose extrapolation from this POD was

performed to derive the OSF. Has the modeling been appropriately conducted and clearly described based on EPA's draft *Benchmark Dose Technical Guidance Document* (U.S. EPA, 2000)? Has the choice of the benchmark response (BMR) for use in deriving the POD (i.e., a BMR of 10% extra risk of the incidence of liver tumors in female mice) been supported and clearly described?

7. EPA has concluded that a nonlinear approach is appropriate for extrapolating cancer risk from male rats to humans because the mode of action analysis suggests that rat bladder tumors occur only after a series of events that begin with calculi formation. At exposure levels below the RfD (i.e., below exposure levels needed to form calculi), no increased risk of cancer is expected. Please comment on whether this approach is scientifically supported and clearly described. Please identify and provide the rationale for any other extrapolation approaches that should be selected.

Inhalation Unit Risk (IUR)

8. The draft Toxicological Review of Biphenyl did not derive an IUR due to the lack of available studies. Are there available data to support the derivation of an IUR for biphenyl? If so, please identify these data.

Charge Example 4: Integrated Science Assessment for a National Ambient Air Quality Standards HISA

UNITED STATES ENVIRONMENTAL PROTECTION AGENCY
NATIONAL CENTER FOR ENVIRONMENTAL ASSESSMENT
WASHINGTON, DC 20460

December 6, 2013

OFFICE OF
RESEARCH AND DEVELOPMENT

SUBJECT: CASAC Review of First External Review Draft Integrated Science Assessment for Oxides of
Nitrogen - Health Criteria

FROM: John Vandenberg, Ph.D.
Director
National Center for Environmental Assessment
Research Triangle Park Division (B243-01)

TO: Aaron Yeow, M.P.H.
Designated Federal Officer
Clean Air Scientific Advisory Committee
EPA Science Advisory Board Staff Office (1400R)

The First *External Review Draft Integrated Science Assessment* (ISA) for *Oxides of Nitrogen – Health Criteria*
prepared by the Environmental Protection Agency's (EPA) National Center for Environmental Assessment -
Research Triangle Park Division (NCEA-RTP) as part of EPA's ongoing review of the primary (health-based)
national ambient air quality standards (NAAQS) for nitrogen dioxide (NO_2) was released on November 22, 2013.
Electronic copies are available for download at http://www.epa.gov/ncea. The draft ISA will be reviewed by the
Clean Air Scientific Advisory Committee (CASAC) 02 Primary NAAQS Review Panel at a public meeting to be
held March 12-13, 2014. We are in the process of distributing the draft ISA for Oxides of Nitrogen to the CA
SAC Oxides of Nitrogen Panel. I am requesting that you forward our charge to the CASAC Oxides of Nitrogen
Panel.

The purpose of the draft ISA is to identify, evaluate, and summarize scientific information on the health effects
associated with gaseous oxides of nitrogen. The ISA is intended to "accurately reflect the latest scientific
knowledge useful in indicating the kind and extent of identifiable effect s on public health which may be expected
from the presence of [a] pollutant in ambient air" (Clean Air Act, Section 108; 42 U.S.C. 7408). This first external
review draft ISA integrates the scientific evidence for review of the primary (health-based) NAAQ S for NO_2 and
pro vides draft findings, conclusions, and judgment s on the strength, coherence, and plausibility of the evidence.
The Preamble presents the process for ISA development, including aspects considered in judging the overall
weight of evidence and framework for causal determination. Criteria used to identify relevant studies for inclusion
in the ISA are also described in the Preamble. Chapter I provides an integrative summary and conclusions of this
assessment. This chapter is supported by detailed information on the relevant evidence available from the multiple
disciplines and approaches related to the causal framework (Preamble to the ISA); atmospheric chemistry,
ambient concentration s, and exposure to oxides of nitrogen (Chapter 2); dosimetry and m odes of act ion
(Chapter 3); health effects of short term exposure to oxides of nitrogen (Chapter 4); health effects of long- term
exposure to oxides of nitrogen (Chapter 5); and lifestages and populations potentially at increased for health
effects related to oxides of nitrogen (Chapter 6). The final ISA for Oxides of Nitrogen, in conjunction with
additional technical assessments, will provide the scientific basis for EPA's decision regarding the adequacy of
the primary NAAQS for NO_2 to protect human health.

The purpose of this memo is to provide charge questions related to a number of important topics addressed in the ISA. Following the CASAC and public review of the draft ISA, NCEA-RTP will produce a second draft ISA, which will be released the summer of 2014.

Charge to the CASAC Oxides of Nitrogen Panel

EPA has aimed to succinctly present and integrate the policy-relevant scientific evidence for the review of the NO_2 NAAQS while also sufficiently describing how scientific information was evaluated in forming the conclusions presented. Previous panels have emphasized the importance of older studies and concluded that if older studies are open to reinterpretation in light of newer data and/or they remain the definitive works available in the literature, they should be discussed in detail to reinforce key concepts and conclusions. In considering subsequent charge questions and recognizing an overall goal of producing a clear and concise document, are there topics that should be added or receive additional discussion? Similarly, are there topics for which discussion should be shortened or removed? Does the Panel have opinions on how the document can be shortened without eliminating important and necessary content?

In addition, we ask the Panel to focus on the following specific questions in their review:

1. The Executive Summary is intended to provide a concise synopsis of the key findings and conclusions of the ISA for a broad range of audiences. Please comment on the clarity with which the Executive Summary communicates the key information from the ISA. Please provide recommendation on information that should be added or information that should be left for discussion in the subsequent chapters of the ISA.

2. Chapter 1 summarizes key information from the Preamble about the process for developing an ISA. Chapter 1 also presents the integrative summary and conclusions from the subsequent detailed chapters of the ISA for Oxides of Nitrogen and characterizes available scientific information on policy-relevant issues.

   a. Please comment on the usefulness and effectiveness of the summary presentation. Please provide recommendations on approaches that may improve the communication of key ISA findings to varied audiences and the synthesis of available information across subject areas.

   b. What are the Panel's thoughts on the application of the Health and Environmental Research Online (HERO) system to support a more transparent assessment process?

   c. To what extent does Chapter 1 communicate the key scientific information on sources, atmospheric chemistry, ambient concentrations, exposure, and health effects of oxides of nitrogen as well as at-risk lifestages and populations? What information should be added or is more appropriate to leave for discussion in the subsequent detailed chapters?

   d. What are the Panel's thoughts on the rationale presented for forming causal determinations for NO2 exposure only and considering epidemiologic results for associations between NOX and health effects in causal determinations for NO2 (Sections 1.4.1 and 1.4.3)?

e. Based on individual Panel member recommendations from June 2013[1] on the *Draft Plan for the Development of the Integrated Science Assessment for Nitrogen Oxides – Health Criteria* (May 2013)[2], Chapter 1 presents an integrated evaluation of various epidemiologic lines of evidence that inform the independent effects of NO2 exposure (Section 1.5). This section discusses available information that is not necessarily included in the health effect chapters on potential confounding by copollutants and other factors as well as the potential for NO2 to serve primarily as an indicator of traffic-related pollutants and traffic proximity. This discussion is in Chapter 1 because it integrates information across Chapters 2, 4, and 5. Please comment on the extent to which this discussion is informative in describing how the evidence of independent effects of NO2 is evaluated in this ISA. Does the discussion accurately reflect the available evidence? If this discussion is informative, what information could be added or removed to improve the discussion. Should the discussion remain in Chapter 1 or should it be moved to another part of the ISA?

f. Please comment on the extent to which the discussion of various policy-relevant considerations is clearly described and integrates relevant information (Section 1.6). Please identify any other relevant information that would be useful to include.

3. Chapter 2 describes scientific information on sources, atmospheric chemistry, air quality characterization, and human exposure of oxides of nitrogen.

a. To what extent is the information presented regarding characteristics of sources, chemistry, monitoring concentrations, and human exposure accurate, complete, and relevant to the review of the NO_2 NAAQS?

b. To what extent are the analyses of air quality presented clearly conveyed, appropriately characterized, and relevant to the review of the NO_2 NAAQS?

c. How effective are the source category groupings and the discussion of source emissions in understanding the importance and impacts of oxides of nitrogen from different sources on both national and local scales?

d. Please comment on the extent to which available information on the spatial and temporal trends of ambient oxides of nitrogen at various scales has been adequately and accurately described.

e. Please comment on the accuracy, level of detail, and completeness of the discussion regarding exposure assessment and the influence of exposure error on effect estimates in epidemiologic studies of the health effects of NO_2.

4. Chapter 3 characterizes scientific evidence on the dosimetry and modes of action for NO_2 and nitric oxide (NO). Dosimetry and modes of action are bridged by reactions of NO_2 with components of the extracellular lining fluid and by reactions of NO with heme proteins, processes that play roles in both uptake and biological responses.

a. Given the ubiquity of reactive substrates and reaction rate of NO_2 with these substrates, it appears unlikely NO_2 itself will penetrate through the lung lining fluid to the epithelium (see Table 3-1). Please comment of the adequacy of the discussion of NO_2 uptake and reactivity in the respiratory tract.

---

[1] The individual panel member comments are available at
http://yosemite.epa.gov/sab/sabproduct.nsf/08EF0A3789CDB13A85257B8E006A496E/$File/EPA-CASAC-13-006+unsigned.pdf
[2] The draft plan for development of the ISA is available at
http://yosemite.epa.gov/sab/sabproduct.nsf/4620a620d0120f93852572410080d786/bc264e65792e015f85257b4a0 07128c6!OpenDocument

b. Since existing dosimetric models for NO_2 do not consider the probability of oxidants/cytotoxic products reaching target sites, it was concluded that these models are inadequate for within or cross species comparisons. Please comment on the validity of this conclusion and identify and comment on the validity of any alternative conclusions.

c. Please comment on the adequacy of the discussion of endogenously occurring NO_2 and NO and their reaction products in comparison to that derived from ambient inhalation.

d. To what extent are the discussion and integration of the potential modes of action underlying the health effects of exposure to oxides of nitrogen presented accurately and in sufficient detail? Are there additional modes of action that should be included in order to characterize fully the underlying mechanisms of oxides of nitrogen?

5. Chapters 4 and 5 present assessments of the health effects associated with short-term and long-term exposure to oxides of nitrogen, respectively. The discussion is organized by health effect category, outcome, and scientific discipline.

a. To what extent do the discussions in this chapter accurately reflect the body of evidence from epidemiologic, controlled human exposure and toxicological studies?

b. Please comment on the balance of discussion of evidence from previous and recent studies in informing the causal determinations.

c. Please comment on the adequacy of the discussion of the strengths and limitations of the evidence in the text and tables within Chapters 4 and 5 and in the evaluation of the evidence in the causal determinations.

d. What are the views of the panel on the integration of epidemiologic, controlled human exposure, and toxicological evidence, in particular, on the balance of emphasis placed on each source of evidence? Please comment on the adequacy with which issues related to exposure assessment and mode of action are integrated in the health effects discussion. Please provide recommendations on information in other chapters of the ISA that would be useful to integrate with the health effects discussions in these chapters.

e. Please comment on the appropriateness of using experimental and epidemiologic evidence for morbidity effects to inform the biological plausibility of total mortality associated with short-term (Section 4.4) and long-term (Section 5.5) NO_2 exposure and in turn, to inform causal determinations.

f. Section 4.2.2 discusses the effect of short-term NO_2 exposure on airways responsiveness. This section focuses primarily on an EPA meta-analysis developed for this ISA of airway responsiveness data for individuals with asthma and secondarily on the potential of various factors to affect airways hyperresponsiveness independently or in conjunction with NO_2 exposure in controlled human exposure studies. This material presently is unpublished and we ask the Panel to provide the peer review for the analysis, in particular, to comment on the appropriateness of the methodology utilized for the meta-analysis, the conclusions reached based this analysis, and its use in the draft ISA. With regard to factors potentially affecting airways responsiveness, please comment on the adequacy of this discussion. Are there other modifying factors that should be considered?

g. The 2008 ISA for Oxides of Nitrogen stated that one of the largest uncertainties was the potential for health effects observed in association with NO_2 exposure to be confounded by correlated copollutants. To what extent has evidence that informs independent effects of NO_2 been adequately discussed in Chapters 4 and 5 and appropriately interpreted as reducing uncertainty (for example, evaluation of copollutant model results)? Has the current draft ISA appropriately considered recent epidemiologic findings regarding potential copollutant confounding in causal determinations? Please provide comments specifically for respiratory effects, cardiovascular effects, and total mortality of short-term NO_2 exposure.

h. To what extent is the causal framework transparently applied to evidence for each of the health effect categories evaluated to form causal determinations? How consistently was the causal framework applied across the health effect categories? Do the text and tables in the summaries and causal determinations clearly communicate how the evidence was considered to form causal determinations?

i. What are the views of the panel regarding the clarity and effectiveness of figures and tables in conveying information about the consistency of evidence for a given health endpoint? In particular, was the use of the tables and figures in both the text and online in the HERO database effective in providing additional information on the studies evaluated? Are there tables and figures in the ISA that would be more appropriate to include as a resource in the HERO database?

6. Chapter 6 evaluates scientific information and presents conclusions on factors that may modify exposure to NO_2, physiological responses to NO_2 exposure, or risk of health effects associated with NO_2 exposure. Consistent with the ISAs for ozone and lead, conclusions on these at-risk factors inform at-risk lifestages and populations.

a. How effective are the categories of at-risk factors in providing information on potential at-risk lifestages and populations? Is there information available on other key at-risk factors that is not included in the first draft ISA and should be added?

b. To what extent do the discussions in this chapter accurately reflect the body of available evidence from epidemiologic, controlled human exposure, and toxicological studies, including the extent to which evidence indicates that the effects of NO_2 exposure are independent of other traffic-related copollutants?

c. Please comment on the consistency and transparency with which the framework for drawing conclusions about at-risk factors has been applied in this ISA.

d. To what extent is available scientific evidence on factors that modify exposure to NO_2 discussed in the chapter and adequately considered in conclusions for at-risk lifestages or populations?

We look forward to discussing these issues with the CASAC Oxides of Nitrogen Panel at our upcoming meeting. Should you have any questions regarding the draft ISA for Oxides of Nitrogen, please feel free to contact Dr. Steven Dutton (919-541-5035, dutton.steven@epa.gov) or Dr. Molini Patel (919-541-1492, patel.molini@epa.gov).

cc: Aaron Yeow, SAB, OA
Kenneth Olden, ORD/NCEA
Reeder Sams, ORD/NCEA
Steven Dutton, ORD/NCEA
Molini Patel, ORD/NCEA
Mary Ross, ORD/NCEA
Deirdre Murphy, OAR/OAQPS
Erika Sasser, OAR/OAQPS
Beth Hassett-Sipple, OAR/OAQPS

Charge Example 5: Science Advisory Board Example

**Animal Feeding Operations Air Emissions Estimating Methodologies
From the National Air Emissions Monitoring Study**

MEMORANDUM

This memorandum requests that the Science Advisory Board (SAB) review and comment on the draft emissions estimating methodologies (EEMs) for animal feeding operations (AFOs). In preparation for this review, the SAB has formed the *Animal Feeding Operations Emission Review Panel*. We envision conducting multiple meetings of this panel to cover the material we are requesting to be reviewed. This memorandum contains background material and charge questions for review by the expert SAB Panel at the initial meeting. We request that these materials be forwarded to the SAB Panel for their review.

As the attachment and associated documents illustrate, the EPA staff has carefully considered the data collected as part of the National Air Emissions Monitoring Study (NAEMS) and now ask the Panel to refine and comment upon our work thus far to create EEMs. To bound and define the discussion, the attachment offers charge questions for the Panel to consider.

By way of background, in 2005, the EPA entered a voluntary consent agreement with the AFO industry in which AFOs that chose to sign the Air Compliance Agreement (Agreement) shared responsibility for funding a nationwide emissions monitoring study. The NAEMS monitoring protocol was developed through a collaborative effort of AFO industry experts, university scientists, U.S. Department of Agriculture and EPA scientists and other stakeholders. The monitoring study was designed to gather data for developing methodologies for estimating emissions from AFOs and to help AFOs determine and comply with their regulatory responsibilities under the Clean Air Act (CAA), the Comprehensive Environmental Response, Compensation and Liability Act (CERCLA), and the Emergency Planning and Community Right-To-Know Act (EPCRA). Once the EPA publishes the applicable EEMs, the Agreement requires each participating AFO to certify that it is in compliance with all relevant requirements of the CAA, CERCLA and EPCRA.

We appreciate your efforts and those of the Panel to prepare for the upcoming meeting and look forward to discussing this project in detail. Questions regarding the attached materials should be directed to [name], EPA-OAQPS ([telephone]; [email]).

Attachment

ATTACHMENT

Regulatory Background

In 2005, the EPA entered a voluntary consent agreement with the animal feeding operations (AFO) industry in which AFOs that chose to sign the Air Compliance Agreement (Agreement) shared responsibility for funding the National Air Emissions Monitoring Study (NAEMS). Approximately 2,600 AFOs, representing nearly 14,000 facilities that include broiler, dairy, egg layer and swine operations, received the EPA's approval to participate in the Agreement.

To provide a framework for the NAEMS, AFO industry experts, university and government scientists and other stakeholders collaborated to develop a comprehensive monitoring plan. The study was designed to generate scientifically credible data to characterize emissions from the participating animal sectors.

Consistent with the Agreement, the Agriculture Air Research Council (AARC), a nonprofit entity comprised of participating AFO industry representatives, administered the monitoring study. The AARC was responsible for selecting the Independent Monitoring Contractor (IMC) and the study's Science Advisor with EPA approval. The Agreement outlined the roles and responsibilities of the AARC, the IMC and the Science Advisor.

The monitoring plan specified the general geographic location of the farms to be monitored, animal production phase, ventilation type, manure management/handling system and other pertinent information for each animal sector.

- For broilers, two sites were to be monitored - one on the West Coast and the other in the Southeast. Both were to be mechanically ventilated and have litter on the floor.

- For the swine industry, the sites were to be located in the Southeast (sow and finisher), Midwest (sow and finisher), and West (sow). Mechanically-ventilated buildings, a deep pit building, lagoons and basin manure storage types were to be monitored.

- For dairy, both naturally- and mechanically-ventilated buildings, lagoons and basins were monitored. Five dairies were monitored, one dairy in each of the following geographical areas: Northeast, Midwest, Northwest, West and South.

For confinement sources, the IMC monitored for ammonia (NH3), particulate matter (PM10, PM2.5, TSP), volatile organic compounds (VOCs) and hydrogen sulfide (H2S). For lagoons and basins, H2S, NH3 and VOC were to be monitored. Accordingly, the EPA is then responsible for developing EEMs for each of these pollutants.

Charge to the Science Advisory Board (SAB) AFO Air Emissions Review Panel

In preparation for the first and second meeting, the EPA has analyzed the NAEMS data for two broiler sites and nine swine and dairy lagoons/basins. For the purpose of this study, the EPA used the description of a lagoon and basin as provided in the MidWest Plan Service "Manure Storages" (MWPS-18 Section 2) document. According to MWPS, "A lagoon is a biological treatment system designed and operated for biodegradation of organic matter in animal manure to a more stable end product. A basin, while similar to but smaller than a lagoon, is designed to store manure only and is not a treatment system."

For a broiler confinement house, the EPA has developed draft EEMs for NH3, PM10, PM2.5, TSP, VOC and H2S. For swine and dairy lagoons/basins, the EPA has only developed a draft EEM for NH3. The documents provided to the SAB describe the sites monitored; the data submitted to the EPA; and a detailed discussion of the statistical methodology used to develop the draft EEMs. This material is provided to inform the SAB panel of the EEM development process used by the agency. In subsequent meetings, the EPA will address draft EEMs for egg-layers, swine and dairy confinement houses and other pollutants for swine and dairy lagoons/basins.

Issue 1: Statistical Methodology used to develop draft EEMs

The EPA seeks the SAB's input on the statistical methodology used by the EPA to develop the draft EEMs. Section 7.0 and 8.0 of the broiler document and Section 5.0 of the swine and dairy lagoon/basin document provide an overview of the statistical methodology used to develop the draft EEMs. A flow diagram of the statistical methodology is provided in Figure 7-1 in the broiler document and Figure 5-1 in the swine and dairy lagoon/basin document. The EPA considers this statistical methodology to be the best approach for analyzing the data and intends to use this same approach to develop draft EEMs for the egg-layers, swine and dairy confinement houses.

Using the process described in the sections listed above, we developed a mean trend function that provides a point prediction of emissions under a given set of conditions. We chose an appropriate mean trend function to quantify the relationship between predictor variables and pollutant emissions by analyzing the emissions data and incorporating knowledge of the emissions generating processes. The EEM development process also involves choosing a probability distribution and covariance function to appropriately quantify other contributions to variability in emissions, and thereby to accurately quantify methods at all stages. If necessary, we will adjust the statistical methodology based on our review of the SAB's input.

Question 1: Please comment on the statistical approach used by the EPA for developing the draft EEMs for broiler confinement houses and swine and dairy lagoons/basins. In addition, please comment on using this approach for developing draft EEMs for egg-layers, swine and dairy confinement houses.

Issue 2: Statistical Methodology used to develop swine and dairy lagoon/basin draft EEMs

After conducting an initial analysis of the NAEMS data submitted for swine and dairy lagoons/basins, the EPA decided to focus on developing a draft EEM for NH3. The EPA's review of current literature indicates that lagoon/basin emissions are influenced by several factors, one of these being lagoon/basin temperature. To ensure that the dataset used to develop the draft EEM represented all seasonal meteorological conditions for the entire two year monitoring period, the EPA decided to combine the swine and dairy data. Combining the swine and dairy lagoon/basin dataset also resulted in combining lagoon and basin emissions data.

To maximize the number of NH3 emissions measurements used to develop the draft EEM, the EPA used static predictor variables (SPVs) as surrogates for data on lagoon/basin conditions (i.e., nitrogen content of lagoon liquid, lagoon pH, oxidation reduction potential and temperature). The static variables of animal type, total live mass of animal capacity on the farm and the surface area of the lagoon were used to represent NH3 precursor loading and the potential for release to the air. Consistent with operating parameters associated with statistical degrees-of-freedom, we concluded that two degrees of freedom was the maximum that the data would credibly allow for inclusion in the developing the draft EEM. As a result, the EPA developed three sets of draft EEMs, using the paired combinations of these static variables (i.e., animal type, surface area, farm size) and the continuous variables representing meteorological conditions (i.e., temperature, atmospheric pressure, humidity, wind speed, solar radiation).

Question 2: Please comment on the agency's decision to combine the swine and dairy dataset to ensure that all seasonal meteorological conditions are represented. In addition, the agency also seeks the SAB's comments on whether the agency should combine lagoon and basin data.

Question 3: Please comment on the agency's decision to use SPVs as surrogates for data on lagoon/basin conditions. Given the uncertainties in that approach, does the SAB recommend that the EPA consider specific alternative approaches for statistically analyzing the data that would allow for the site-specific lagoon liquid characteristics to be used as predictor variables?

Question 4: Does the SAB recommend that EPA consider alternative approaches for developing the draft NH3 EEM that balances the competing needs for a large dataset (to reflect seasonal meteorological conditions) versus incorporating additional site-specific factors that directly affect lagoon emissions. If so, what specific alternative approaches would be appropriate to consider?

Issue 3: Negative and Zero Data

Some emissions measurements were reported to the EPA as either negative or zero emissions values. When developing the draft EEMs, the EPA used the following general approach regarding inclusion of negative and zero emissions values in the data.

- The EPA evaluated whether the negative or zero values represent the variability in emissions measurements due to the means of obtaining the measurements. For example, negative values for a pollutant concentration might result when the concentration of the pollutant falls below the minimum detection limit of a monitor. For all EEM datasets, the EPA included zero values because these values potentially represent instances where the emissions from the source were zero (e.g., a frozen lagoon), or the background and pollutant concentrations from the source were the same. Regarding negative values, in cases where the dataset available to develop draft EEMs was relatively large and the emissions were significantly greater than zero, the EPA excluded negative emissions values from the EEM datasets. The EPA used this approach to develop the entire broiler confinement house draft EEMs and swine and dairy lagoon/basin NH3 draft EEMs.

- The EPA reviewed the data to see if the data quality measures were properly performed according to the Quality Assurance Project Plan.

- If the EPA identified data where the quality assurance measures were not followed, we contacted the science advisor to determine if the corrected data could be submitted to the EPA.

The EPA has conducted a preliminary analysis of the swine and dairy lagoon/basin H2S emissions data. Our analysis indicates that we may need to modify our approach for handling negative and zero data in order to develop a draft H2S EEM for swine and dairy lagoons/basins. A modification may be needed due to the limited number of H2S emissions values, the presence of a greater percentage of negative emissions values and emissions values that are closer to zero than the NH3 emissions for swine and dairy lagoons/basins. The EPA's concern is that failure to include the negative measurements in the dataset, or setting them equal to zero, would result in an EEM that fails to fully quantify uncertainty around the point prediction of emissions attributable to measurement error.

Question 5: Please comment on the EPA's approach for handling negative or zero emission measurements.

Question 6: In the interest of maximizing the number of available data values for development of the draft H2S EEMs for swine and dairy lagoons/basins, does SAB recommend any alternative approaches for handling negative and zero data other than the approach used by the agency.

Issue 4: Volatile Organic Compounds (VOC) Data

The EPA reviewed the VOC data submitted for the California and Kentucky broiler sites. The two sites used different VOC measurement techniques. Based on our analysis of the measurement and analytical techniques and the VOC data, the EPA decided to use only the VOC data from the Kentucky sites when developing the draft VOC EEM.

Question 7: Please comment on the approach EPA used to develop the draft broiler VOC EEM.

APPENDIX I. EXAMPLES OF *FEDERAL REGISTER* NOTICES REQUESTING PUBLIC COMMENT

Federal Register Notice: Announcement of Public Comment Period for Draft Document

Federal Register, Volume 77 Issue 102 (Friday, May 25, 2012)
[Federal Register Volume 77, Number 102 (Friday, May 25, 2012)]
[Notices]
[Pages 31353-31355]
From the Federal Register Online via the Government Printing Office [http://www.gpo.gov/]
[FR Doc No: 2012-12808]

ENVIRONMENTAL PROTECTION AGENCY

[FRL-9678-3; Docket ID No. EPA-HQ-ORD-2012-0276]

An Assessment of Potential Mining Impacts on Salmon Ecosystems of Bristol Bay, AK

AGENCY: Environmental Protection Agency (EPA).

ACTION: Notice of public comment period.

SUMMARY: The U.S. Environmental Protection Agency (EPA) is announcing a public comment period for the draft document titled, "An Assessment of Potential Mining Impacts on Salmon Ecosystems of Bristol Bay, Alaska" (EPA-910-R-12-004a-d). The document was prepared by the EPA's Region 10 (Pacific Northwest and Alaska), EPA's Office of Water, and EPA's Office of Research and Development. The EPA conducted this assessment to determine the significance of Bristol Bay's ecological resources and evaluate the potential impacts of large-scale mining on these resources. EPA will use the results of this assessment to inform the consideration of options consistent with its role under the Clean Water Act. The assessment is intended to provide a scientific and technical foundation for future decision making; EPA will not address use of its regulatory authority until the assessment becomes final and has made no judgment about whether and how to use that authority at this time.

DATES: The public comment period began Friday, May 18, 2012, and ends Monday, July 23, 2012. Technical comments should be in writing and must be received by EPA by Monday, July 23, 2012.

ADDRESSES: The draft "An Assessment of Potential Mining Impacts on Salmon Ecosystems of Bristol Bay, Alaska" is available primarily via the Internet on the EPA Region 10 Bristol Bay Web site at www.epa.gov/bristolbay as well as on the National Center for Environmental Assessment's Web site under the Recent Additions and the Data and Publications menus at www.epa.gov/ncea. A printed copy of the assessment will be placed at public locations in Bristol Bay and in Anchorage, AK. These locations are listed on the Region 10 Web site. A limited number of paper copies are available from the Information Management Team, NCEA; telephone: 703-347-8561; facsimile: 703-347-8691. If you are requesting a paper copy, please provide your name, your mailing address, and the document title, "An Assessment of Potential Mining Impacts on Salmon Ecosystems of Bristol Bay, Alaska." Please also indicate if a paper copy of the full set of appendices is needed.

Comments on the report may be submitted electronically via http://www.regulations.gov/, by email, by mail, by facsimile, or by hand delivery/courier. Please follow the detailed instructions provided in the SUPPLEMENTARY INFORMATION section of this notice.

FOR FURTHER INFORMATION CONTACT: For information on the public comment period, contact the Office of Environmental Information Docket; telephone: 202-566-1752; facsimile: 202-566-1753; or email: ORD.Docket@epa.gov.

For technical information concerning the report, contact Judy Smith; telephone: 503-326-6994; facsimile: 503-326-3399; or email: r10bristolbay@epa.gov.

SUPPLEMENTARY INFORMATION:

I. Information About the Project/Document

The U.S. Environmental Protection Agency (EPA) conducted this assessment to determine the significance of Bristol Bay's ecological resources and evaluate the potential impacts of large-scale mining on these resources. The EPA will use the results of this assessment to inform the consideration of options consistent with its role under the Clean Water Act. The assessment is intended to provide a scientific and technical foundation for future decision making. The Web site that describes the project is www.epa.gov/bristolbay. This draft document addresses potential impacts to water quality and the salmon fishery that may result from large-scale mining in the Nushagak and Kvichak watersheds of southwest Alaska.

EPA is releasing this draft assessment for the purposes of public comment and peer review. This draft assessment is not final as described in EPA's information quality guidelines, and it does not represent and should not be construed to represent Agency policy or views. EPA utilizes public comments as one means to ensure that science products are complete and accurate. EPA is seeking comments from the public on all aspects of the report, including the scientific and technical information presented in the report, the hypothetical mining scenario used, the data and information used to inform assumptions about mining activities and the evaluations of risk to the fishery, and the potential mitigation measures considered (and effectiveness of those measures). EPA is also specifically seeking any additional data or scientific or technical information about Bristol Bay resources or large-scale mining that should be considered in our evaluation.

EPA will consider any public comments submitted in accordance with this notice when revising the document. After public review and comment, EPA's independent contractor, Versar, Inc., will convene an expert panel for independent external peer review of this draft assessment. The public comment period and external peer review meeting are separate processes that provide opportunities for all interested parties to comment on the assessment. The preferred method to submit comments is through the docket, which is described below. Public meetings will be held in Anchorage, Dillingham, Newhalen, Naknek, Nondalton, and New Stuyahok, AK during the week of June 4-8, 2012. Spoken comments will be accepted at these meetings. The external peer review panel meeting is scheduled to be held in Anchorage, AK on August 7, 8, and 9, 2012. The public will be invited to attend on August 7 and 8, 2012. Further information regarding the external peer review panel meeting will be announced at a later date in the Federal Register.

II. How To Submit Technical Comments to the Docket at http://www.regulations.gov/

Submit your comments, identified by Docket ID No. EPA-HQ-ORD-2012-0276, by one of the following methods:

http://www.regulations.gov/: Follow the on-line instructions for submitting comments.

Email: ORD.Docket@epa.gov. Include the docket number EPA-HQ-ORD-2012-0276 in the subject line of the message.

Fax: 202-566-1753.

Mail: Office of Environmental Information (OEI) Docket (Mail Code: 2822T), Docket EPA-HQ-ORD-2012-0276, U.S. Environmental Protection Agency, 1200 Pennsylvania Ave. NW., Washington, DC 20460. The phone number is 202-566-1752. If you provide comments by mail, please submit one unbound original with pages numbered consecutively, and three copies of the comments. For attachments, provide an index, number pages consecutively with the comments, and submit an unbound original and three copies.

Hand Delivery: The OEI Docket is located in the EPA Headquarters Docket Center, Room 3334, EPA West Building, 1301 Constitution Ave. NW., Washington, DC. The EPA Docket Center Public

Reading Room is open from 8:30 a.m. to 4:30 p.m., Monday through Friday, excluding legal holidays. The telephone number for the Public Reading Room is 202-566-1744. Deliveries are only accepted during the docket's normal hours of operation, and special arrangements should be made for deliveries of boxed information. If you provide comments by hand delivery, please submit one unbound original with pages numbered consecutively, and three copies of the comments. For attachments, provide an index, number pages consecutively with the comments, and submit an unbound original and three copies.

Comment at a public meeting: Spoken comments will be taken at public meetings during June 4-8, 2012. A court reporter will provide a transcription of comments received at the Anchorage and Dillingham meetings for the docket. Audio recording and written notes will be taken for the docket for comments spoken at Naknek, Newhalen, New Stuyahok, and Nondalton.

Instructions: Direct your comments to Docket ID No. EPA-HQ-ORD-2012-0276. Please ensure that your comments are submitted within the specified comment period. Comments received after the closing date will be marked "late," and may only be considered if time permits. It is EPA's policy to include all comments it receives in the public docket without change and to make the comments available on-line at http://www.regulations.gov/, including any personal information provided, unless a comment includes information claimed to be Confidential Business Information (CBI) or other information whose disclosure is restricted by statute. Do not submit information that you consider to be CBI or otherwise protected through http://www.regulations.gov/ or email. The http://www.regulations.gov/ Web site is an "anonymous access" system, which means EPA will not know your identity or contact information unless you provide it in the body of your comment. If you send an email comment directly to EPA without going through http://www.regulations.gov/, your email address will be automatically captured and included as part of the comment that is placed in the public docket and made available on the Internet. If you submit an electronic comment, EPA recommends that you include your name and other contact information in the body of your comment and with any disk or CD-ROM you submit. If EPA cannot read your comments due to technical difficulties and cannot contact you for clarification, EPA may not be able to consider your comments. Electronic files should avoid the use of special characters and any form of encryption and be free of any defects or viruses. For additional information about EPA's public docket, visit the EPA Docket Center homepage at www.epa.gov/epahome/dockets.htm.

Docket: Documents in the docket are listed in the http://www.regulations.gov/--index. Although listed in the index, some information is not publicly available, e.g., CBI or other information whose disclosure is restricted by statute. Certain other material, such as copyrighted material, will be publicly available only in hard copy.

Publicly available docket materials are available either electronically at http://www.regulations.gov/ or in hard copy at the OEI Docket in the EPA Headquarters Docket Center.

Dated: May 21, 2012.
Darrell Winner,
Acting Director, National Center for Environmental Assessment.
[FR Doc. 2012-12808 Filed 5-24-12; 8:45 am]
BILLING CODE 6560-50-P

Federal Register Notice: Announcement of Peer Review Panel Members and Public Comment Period for Draft Charge Questions

Federal Register, Volume 77 Issue 108 (Tuesday, June 5, 2012)
[Federal Register Volume 77, Number 108 (Tuesday, June 5, 2012)]
[Notices]
[Pages 33213-33215]
From the Federal Register Online via the Government Printing Office [http://www.gpo.gov/]
[FR Doc No: 2012-13431]

===

ENVIRONMENTAL PROTECTION AGENCY

[FRL-9681-3; EPA-HQ-ORD-2012-0358]

An Assessment of Potential Mining Impacts on Salmon Ecosystems of
Bristol Bay, Alaska--Peer Review Panel Members and Charge Questions

AGENCY: Environmental Protection Agency (EPA).

[[Page 33214]]

ACTION: Notice of availability and public comment period.

SUMMARY: EPA is announcing the peer review panel members assembled by
an independent contractor to evaluate the draft document titled, "An
Assessment of Potential Mining Impacts on Salmon Ecosystems of Bristol
Bay, Alaska" (EPA-910-R-12-004a-c). EPA is also announcing a three
week public comment period for the draft charge questions to be
provided to the peer review panel. The assessment was prepared by the
U.S. EPA's Region 10 Office (Pacific Northwest and Alaska), EPA's
Office of Water, and EPA's Office of Research and Development. The U.S.
EPA conducted this assessment to determine the significance of Bristol
Bay's ecological resources and evaluate the potential impacts of large-
scale mining on these resources.

DATES: The public comment period begins June 5, 2012, and ends June 26,
2012. Comments should be in writing and must be received by EPA by June
26, 2012.

Availability: Draft charge questions are provided below. Copies of the draft charge questions are also available via the Internet on the EPA Region 10 Bristol Bay Web site at www.epa.gov/bristolbay. The draft document "An Assessment of Potential Mining Impacts on Salmon Ecosystems of Bristol Bay, Alaska" is also available on the Internet on the EPA Region 10 Bristol Bay Web site at www.epa.gov/bristolbay. A limited number of paper copies of the draft charge questions are available from the Information Management Team, NCEA; telephone: 703-347-8561; facsimile: 703-347-8691. If you are requesting a paper copy, please provide your name, your mailing address, and title, "Peer Review Charge Questions on An Assessment of Potential Mining Impacts on Salmon Ecosystems of Bristol Bay, Alaska."

Comments on the draft charge questions may be submitted electronically via http://www.regulations.gov/, by email, by mail, by facsimile, or by hand delivery/courier. Please follow the detailed instructions provided in the SUPPLEMENTARY INFORMATION section of this notice.

FOR FURTHER INFORMATION CONTACT: For information on the public comment period, contact the Office of Environmental Information Docket; telephone: 202-566-1752; facsimile: 202-566-9744; or email: ORD.Docket@epa.gov.

For technical information concerning the report, contact Judy Smith; telephone: 503-326-6994; facsimile: 503-326-3399; or email: r10bristolbay@epa.gov.

SUPPLEMENTARY INFORMATION:

I. Information About the Project

The U.S. EPA conducted this assessment to determine the significance of Bristol Bay's ecological resources and evaluate the potential impacts of large-scale mining on these resources. The U.S. EPA will use the results of this assessment to inform the consideration of options consistent with its role under the Clean Water Act. The assessment is intended to provide a scientific and technical foundation for future decision making. The Web site that describes the project is www.epa.gov/bristolbay.

EPA released the draft assessment for the purposes of public comment and peer review on May 18, 2012. Consistent with guidelines for the peer review of highly influential scientific assessments, EPA asked a contractor (Versar, Inc.) to assemble a panel of experts to evaluate the draft report. Versar evaluated the 86 candidates nominated during a previous public comment period (February 24, 2012 to March 16, 2012) and sought other experts to complete this peer review panel. The twelve peer review panel members are as follows:

Mr. David Atkins, Watershed Environmental, LLC.--Expertise in mining and hydrology.

Mr. Steve Buckley, WHPacific/NANA Alaska--Expertise in mining and seismology.

Dr. Courtney Carothers--Expertise in indigenous Alaskan cultures.

Dr. Dennis Dauble, Washington State University--Expertise in fisheries biology and wildlife ecology.

Dr. Gordon Reeves, USDA Pacific NW Research Station--Expertise in fisheries biology and aquatic biology.

Dr. Charles Slaughter, University of Idaho--Expertise in hydrology.

Dr. John Stednick, Colorado State University--Expertise in hydrology and biogeochemistry.

Dr. Roy Stein, Ohio State University--Expertise in fisheries and aquatic biology.

Dr. William Stubblefield, Oregon State University--Expertise in aquatic biology and ecotoxicology.

Dr. Dirk van Zyl, University of British Columbia--Expertise in mining and biogeochemistry.

Dr. Phyllis Weber Scannel--Expertise in aquatic ecology and ecotoxicology.

Dr. Paul Whitney--Expertise in wildlife ecology and ecotoxicology.

The peer review panel will be provided with draft charge questions to guide their evaluation of the draft assessment. These draft charge questions are designed to focus reviewers on specific aspects of the report. EPA is seeking comments from the public on the draft charge questions and welcome input on additional charge questions consistent with the objectives of the assessment. The draft charge questions are as follows:

(1) The assessment brought together information to characterize the ecological, geological, and cultural resources of the Nushagak and Kvichak watersheds. Was this characterization accurate? Was any significant literature missed that would be useful to complete this characterization?

(2) A formal mine plan or application is not available for the porphyry copper deposits in the Bristol Bay watershed. EPA developed a hypothetical mine scenario for its risk assessment. Given the type and location of copper deposits in the watershed, was this hypothetical mine scenario realistic? Has EPA appropriately bounded the magnitude of potential mine activities with the minimum and maximum mine sizes used in the scenario? Is there significant literature not referenced that would be useful to refine the mine scenario?

(3) EPA assumed two potential modes for mining operations: A no-failure mode of operation and a mode outlining one or more types of failures. The no-failure operation mode assumes best practical engineering and mitigation practices are in place and in optimal operating condition. Is the no-failure mode of operation adequately described? Is the choice of engineering and mitigation practices reasonable and consistent with current practices?

(4) Are the potential risks to salmonid fish due to habitat loss and modification and water quantity/quality changes appropriately characterized and described for the no-failure mode of operation? Does the assessment appropriately describe the risks to salmonid fish due to

operation of a transportation corridor under the no-failure mode of operation?

(5) Do the failures outlined in the assessment reasonably represent potential system failures that could occur at a mine of the type and size outlined in the mine scenario? Is there a significant type of failure that is not described? Are the assumed risks of failures appropriate?

(6) Does the assessment appropriately characterize risks to salmonid fish due to a potential failure of water and leachate collection and treatment from the mine site? If not, what suggestions do you have for improving this part of the assessment?

(7) Does the assessment appropriately characterize risks to salmonid fish due to culvert failures along the transportation corridor? If not, what suggestions do you have for improving this part of the assessment?

(8) Does the assessment appropriately characterize risks to salmonid fish due to pipeline failures? If not, what suggestions do you have for improving this part of the assessment?

(9) Does the assessment appropriately characterize risks to salmonid fish due to a potential tailings dam failure? If not, what suggestions do you have for improving this part of the assessment?

(10) Does the assessment appropriately characterize risks to wildlife and human cultures due to risks to fish? If not, what suggestions do you have for improving this part of the assessment?

(11) Does the assessment appropriately describe the potential for cumulative risk from multiple mines?

(12) Does the assessment identify the uncertainties and limitations associated with the mine scenario and the identified risks?

The preferred method to submit comments on the draft peer review charge is through the docket, which is described below. This docket is separate from the docket collecting public comments on the draft assessment itself. The EPA will evaluate comments received on these draft charge questions. Charge questions will be finalized and provided to EPA's independent contractor, Versar, Inc., who will convene the expert panel for independent external peer review.

The external peer review panel meeting is scheduled to be held in Anchorage, AK on August 7, 8, and 9, 2012. The public will be invited to attend on August 7 and 8, 2012. Further information regarding the external peer review panel meeting will be announced at a later date in the Federal Register.

II. How to Submit Technical Comments to the Docket at http://www.regulations.gov/

Submit your comments, identified by Docket ID No. EPA-HQ-ORD-2012-0358, by one of the following methods:
http://www.regulations.gov/: Follow the on-line instructions for

submitting comments.

Email: ORD.Docket@epa.gov. Include the docket number EPA-HQ-ORD-2012-0358 in the subject line of the message.

Fax: 202-566-9744.

Mail: Office of Environmental Information (OEI) Docket (Mail Code: 28221T), Docket EPA-HQ-ORD-2012-0358, U.S. Environmental Protection Agency, 1200 Pennsylvania Avenue NW., Washington, DC 20460. The phone number is 202-566-1752. If you provide comments by mail, please submit one unbound original with pages numbered consecutively, and three copies of the comments. For attachments, provide an index, number pages consecutively with the comments, and submit an unbound original and three copies.

Hand Delivery: The OEI Docket is located in the EPA Headquarters Docket Center, Room 3334, EPA West Building, 1301 Constitution Avenue NW., Washington, DC. The EPA Docket Center Public Reading Room is open from 8:30 a.m. to 4:30 p.m., Monday through Friday, excluding legal holidays. The telephone number for the Public Reading Room is 202-566-1744. Deliveries are only accepted during the docket's normal hours of operation, and special arrangements should be made for deliveries of boxed information. If you provide comments by hand delivery, please submit one unbound original with pages numbered consecutively, and three copies of the comments. For attachments, provide an index, number pages consecutively with the comments, and submit an unbound original and three copies.

Instructions: Direct your comments to Docket ID No. EPA-HQ-ORD-2012-0358. Please ensure that your comments are submitted within the specified comment period. Comments received after the closing date will be marked "late," and may only be considered if time permits. It is EPA's policy to include all comments it receives in the public docket without change and to make the comments available online at http://www.regulations.gov/, including any personal information provided, unless a comment includes information claimed to be Confidential

Business Information (CBI) or other information whose disclosure is restricted by statute. Do not submit information that you consider to be CBI or otherwise protected through http://www.regulations.gov/ or email. The http://www.regulations.gov/ Web site is an "anonymous access" system, which means EPA will not know your identity or contact information unless you provide it in the body of your comment. If you send an email comment directly to EPA without going through http://www.regulations.gov/, your email address will be automatically captured and included as part of the comment that is placed in the public docket and made available on the Internet. If you submit an electronic comment, EPA recommends that you include your name and other contact information in the body of your comment and with any disk or CD-ROM you submit. If EPA cannot read your comments due to technical difficulties and cannot contact you for clarification, EPA may not be able to consider your comments. Electronic files should avoid the use of special characters and any form of encryption and be free of any defects or viruses. For additional information about EPA's public docket visit the EPA Docket Center homepage at www.epa.gov/epahome/dockets.htm.

Docket: Documents in the docket are listed in the http://www.regulations.gov. Although listed in the index, some information is not publicly available, e.g., CBI or other information whose disclosure is restricted by statute. Certain other material, such as copyrighted material, will be publicly available only in hard copy. Publicly available docket materials are available either electronically in http://www.regulations.gov/ or in hard copy at the OEI Docket in the EPA Headquarters Docket Center.

Dated: May 30, 2012.
Darrel A. Winner,
Acting Director, National Center for Environmental Assessment.
[FR Doc. 2012-13431 Filed 6-4-12; 8:45 am]
BILLING CODE 6560-50-P

Federal Register Notice: Peer Review Meeting Announcement and Invitation to Public to Attend and Offer Testimony

Federal Register, Volume 77 Issue 130 (Friday, July 6, 2012)
[Federal Register Volume 77, Number 130 (Friday, July 6, 2012)]
[Notices]
[Pages 40037-40039]
From the Federal Register Online via the Government Printing Office [http://www.gpo.gov/]
[FR Doc No: 2012-16441]

ENVIRONMENTAL PROTECTION AGENCY

[FRL-9697-3]

Notice of the Peer Review Meeting for EPA's Draft Report Entitled
An Assessment of Potential Mining Impacts on Salmon Ecosystems of
Bristol Bay, AK

AGENCY: U.S. Environmental Protection Agency (EPA).

ACTION: Notice of external peer review meeting.

SUMMARY: The U.S. Environmental Protection Agency (EPA) is announcing
that Versar, Inc., an EPA contractor for external peer review, has
convened a panel of experts and will organize and conduct an
independent expert external peer review meeting on August 7-9, 2012, to
review the draft report entitled An Assessment of Potential Mining
Impacts on Salmon Ecosystems of Bristol Bay, Alaska. Versar, Inc.
invites the public to register to attend the first two days of this
meeting as observers. In addition, Versar, Inc. invites the public to
register to provide oral testimony during Day 1 (August 7, 2012)
of the external peer review meeting. The panel will meet privately
on Day 3 (August 9, 2012) of the meeting. The expert panel
is charged with reviewing the scientific and technical
merit of the draft assessment. The panel will not be making
recommendations to the EPA concerning any potential future actions or
policies. Therefore, the peer review meeting will focus on issues of
science relevant to the assessment, rather than its policy
implications. The panel will have access to public comments received in

the official public docket (docket ID number EPA-HQ-ORD-2012-0276) during the assessment's public comment period, as well as oral comments made on Day 1 of the peer review meeting. The draft assessment is available through http://www.regulations.gov/ and at www.epa.gov/bristolbay. In preparing the final assessment, EPA will consider Versar, Inc.'s report of the comments and recommendations from the external peer review meeting, as well as written public comments received through the official public docket. The final peer review report prepared by Versar, Inc. will be made available to the public. EPA has released this draft assessment for the purposes of public comment and peer review. This draft assessment is not final as described in EPA's information quality guidelines, and it does not represent and should not be construed to represent Agency policy or views.

DATES: The public peer review panel meeting will be held on August 7-8, 2012, beginning and ending at approximately 8:30 a.m. and 5:00 p.m. (AKDT) on both days.

ADDRESSES: The independent expert external peer review meeting will be held at the Dena'ina Civic & Convention Center, located at 600 West Seventh Avenue, Anchorage, Alaska.

Meeting Background: As part of the peer review process for the EPA's draft assessment report, the public portion of the peer review meeting will be held on August 7-8, 2012 at the Dena'ina Civic & Convention Center in Anchorage, Alaska. On both days, the meeting will begin at 8:30 a.m. (AKDT) and will end at approximately 5:00 p.m. (AKDT). Members of the public and any other interested parties may register to attend both days of the meeting as observers, and to offer oral testimony on the first day of the meeting.

The focus of this peer review meeting is the scientific content and merit of the EPA's draft assessment. Public speakers are encouraged to focus on issues directly relevant to science-based aspects of the assessment, and to address specific scientific points in their oral testimony. The peer review process is separate from the EPA public comment meetings held in early June that enabled members of the public to provide comments and voice opinions concerning the EPA's draft assessment report and its potential policy implications for the public docket.

Day 1 of the meeting (August 7, 2012) will be dedicated to hearing oral comments on the draft assessment. Members of the public who have registered in advance to provide oral comments will have the opportunity to speak during the observer comment session. Each speaker will be allowed between 3-5 minutes, depending on number of speakers registered. Given time constraints, a maximum of 100 speakers will be allowed to offer testimony. If more than 100 speakers register to provide oral comments, speakers will be selected by Versar in a manner designed to optimize representation from all organizations, affiliations, and present a balance of science issues relevant to the Agency's science assessment. Additional information on selection of speakers and speaking times will be sent out by August 3, to all individuals who register to speak.

To accommodate as many speakers as possible, registered speakers will present oral comments only, without visual aids or written material. All members of the public, including registered observers and speakers, are encouraged to submit written comments and materials to the official public docket for the draft assessment (docket ID number EPA-HQ-ORD- 2012-0276) by the close of the public comment period on July 23, 2012. Panel members will have access to any written comments and materials submitted to the official public docket by this deadline. Registered observers and speakers will not be allowed to distribute any written materials directly to the peer review panel. To submit written comments, please follow one of the methods outlined in the previous Federal Register notice, issued on May 25, 2012, initiating the assessment's public comment period: Federal Register Volume 77, Number 102 (http://www.gpo.gov/fdsys/pkg/FR-2012-05-25/html/2012-12808.htm).

Day 2 of the meeting (August 8, 2012) will be devoted to deliberations of the EPA's draft assessment by the peer review panel, guided by the charge questions provided to the public for public comment. Registered observers may attend and observe the peer review panel deliberations on Day 2, but will not be allowed to address the panel or provide oral or written comments.

Registration: To attend the August 7-8 public portion of the peer review meeting, you must register for the meeting by 11:59 p.m. (EDT) on July 23, 2012. You can register for the meeting by visiting , completing the online registration form, and submitting the required information. You can also register through U.S. Postal Service or overnight/priority mail by sending the necessary registration information (see Required Registration Information) to the Versar Meeting Coordinator, Ms. Brittany Ekstrom, Versar, Inc., 6850 Versar Center, Springfield, VA 22151; Telephone: (703) 642-6767. Registrations sent via U.S. Postal Service or overnight/priority mail must be received by 11:59 p.m. (EDT) on July 23, 2012. There will be no on-site registration, so members of the public who do not register by July 23, 2012 via one of the methods detailed above will not be able to attend the peer review meeting.

Required Registration Information: To register for the meeting online or via post, you must provide your full name, organization or affiliation, and contact information. You must also indicate which days you plan to attend the meeting and if you are interested in making an oral statement during the public comment session on Day 1 of the meeting. If you register to speak, you must also indicate if you have any special requirements related to your oral comments (e.g., translation).

If you indicate that you wish to make oral comments, you will be asked to select one category most closely reflecting the content of your comments. These comment categories are: (i) Mine scenario and operational modes; (ii) potential failures and probabilities; (iii) hydrology; (iv) toxicity; (v) potential effects on Alaska Native culture; (vi) potential effects on fish; (vii) potential effects on wildlife; and (viii) other issues. Should more than 100 speakers register, these categories will be used to ensure that a balance of substantive science issues relevant to the assessment are heard.

FOR FURTHER INFORMATION CONTACT: Questions regarding logistics or registration for the external peer review meeting should be directed to Ms. Brittany Ekstrom, Versar, Inc., 6850 Versar Center, Springfield, VA, 22151; telephone: (703) 642-6767; or via email at BEkstrom@versar.com.

SUPPLEMENTARY INFORMATION:

I. Information About the Project

The EPA conducted this assessment to determine the significance of Bristol Bay's ecological resources and evaluate the potential impacts of large-scale mining on these resources. The EPA will use the results of this assessment to inform the consideration of options consistent with its role under the Clean Water Act. The assessment is intended to provide a sound scientific and technical foundation for future decision making. The Web site that describes the project is www.epa.gov/bristolbay.

II. Information About the Peer Review Panel

The EPA released the draft assessment for the purposes of public comment and peer review on May 18, 2012. Consistent with guidelines for the peer review of highly influential scientific assessments, EPA asked a contractor (Versar, Inc.) to assemble a panel of experts to evaluate the draft report. Versar, Inc. evaluated the 68 candidates nominated during a previous public comment period (February 24, 2012 to March 16, 2012) and sought other experts to complete this peer review panel. The twelve peer review panel members were made public in EPA's previous FRN, issued on June 5, 2012. The panelist's names are included below, with corrections made to account for errors present in the June 5, 2012 FRN:

Mr. David Atkins, Watershed Environmental, LLC.--Expertise in mining and hydrology.
Mr. Steve Buckley, WHPacific--Expertise in mining and seismology.
Dr. Courtney Carothers, University of Alaska Fairbanks--Expertise in indigenous Alaskan cultures.
Dr. Dennis Dauble, Washington State University--Expertise in fisheries

biology and wildlife ecology.

Dr. Gordon Reeves, USDA Pacific NW Research Station--Expertise in fisheries biology and aquatic biology.

Dr. Charles Slaughter, University of Idaho--Expertise in hydrology.

Dr. John Stednick, Colorado State University--Expertise in hydrology and biogeochemistry.

Dr. Roy Stein, Ohio State University--Expertise in fisheries and aquatic biology.

Dr. William Stubblefield, Oregon State University--Expertise in aquatic biology and ecotoxicology.

Dr. Dirk van Zyl, University of British Columbia--Expertise in mining.

Dr. Phyllis Weber Scannell--Expertise in aquatic ecology and ecotoxicology.

Dr. Paul Whitney--Expertise in wildlife ecology and ecotoxicology.

    Dated: June 29, 2012.

Darrell Winner,

Acting Director, National Center for Environmental Assessment.

[FR Doc. 2012-16441 Filed 7-5-12; 8:45 am]

BILLING CODE 6560-50-P

APPENDIX J. CONFLICT OF INTEREST
MEMORANDA FOR ISI

Conflict of Interest Memorandum: Task Orders

U.S. Environmental Protection Agency
Conflict of Interest Statement for Task Orders

The contractor shall include a conflict of interest certification in all task orders in accordance with EPAAR 1552.209-71 and the Section B Clause "Ordering Procedures."

Prior to selecting expert panelists/peer reviewers, the contractor shall perform an evaluation to determine the existence of an actual or potential COI for each potential reviewer. The financial and professional information obtained by the Contractor as part of the evaluation to determine the existence of an actual or potential conflict of interest is considered private and nondisclosable to outside entities except as required by law and/or regulation.

The contractor shall ensure that potential peer reviewers will not have an actual or potential conflict of interest if they are selected to participate in a peer review. When determining if a proposed peer reviewer may have an actual or potential conflict of interest, the contractor shall incorporate the following yes/no questions (a – i) for **all** individuals, and requests for supporting information (j – r) **for task orders involving public peer review meetings** into its established process to evaluate and determine the presence of an actual or potential COI:

Conflict of Interest Analysis		
	YES	NO
a. To the best of your knowledge and belief, is there any connection between the subject topic and any of your and/or your spouse's compensated or uncompensated employment, including government service, during the past 24 months?		
b. To the best of your knowledge and belief, is there any connection between the subject topic and any of your and/or your spouse's research support and project funding, including from any government source, during the past 24 months?		
c. To the best of your knowledge and belief, is there any connection between the subject topic and any consulting by you and/or your spouse, during the past 24 months?		
d. To the best of your knowledge and belief, is there any connection between the subject topic and any expert witness activity by you and/or your spouse, during the past 24 months?		
e. To the best of your knowledge and belief, have you, your spouse, or dependent child, held in the past 24 months, any financial holdings (excluding well-diversified mutual funds and holdings, with a value less than $15,000) with any connection to the subject topic?		
f. Have you made any public statements or taken positions on or closely related to the subject topic under review?		
g. Have you had previous involvement with the development of the document (or review materials) you have been asked to review?		
h. To the best of your knowledge and belief, is there any other information that might reasonably raise a question about an actual or potential personal conflict of interest or bias?		

Conflict of Interest Analysis		
	YES	NO
i. To the best of your knowledge and belief, is there any financial benefit that might be gained by you or your spouse as a result of the outcome of this review?		

Information to be collected from panel members:

j. Compensated and noncompensated employment—for panel member and spouse—sources of compensated and uncompensated employment, including government service, for the preceding 2 years, including a brief description of work.
k. Research funding—for panel member—sources of research support and project funding, including from any government source, for the preceding 2 years for which the panel member served as the Principal Investigator, Significant Collaborator, or Project Manager or Director. For panel member's spouse, a general description of research and project activities in the preceding 2 years.
l. Consulting—for panel member—compensated consulting activities during the preceding 2 years, including names of clients if compensation provided 15% or more of annual compensation. For panel member's spouse, a general description of consulting activities for the preceding 2 years.
m. Expert witness activities—for panel member—sources of compensated expert witness activities and a brief description of issue and testimony. For panel member's spouse, a general description of expert testimony provided in the preceding 2 years.
n. Assets: Stocks, Bonds, Real Estate, Business, Patents, Trademarks and Royalties—for panel member, spouse and dependent children—specific financial holdings that collectively had a fair market value greater than $15,000 at any time during the preceding 24-month period (excluding well-diversified mutual funds, money market funds, treasury bonds and personal residence).
o. Liabilities—for panel member, spouse and dependent children—liabilities over $10,000 owed at any time in the preceding 12 months (excluding a mortgage on personal residence, home equity loans, automobile and consumer loans).
p. Public statements—a brief description of public statements and/or positions on, or closely related to, the matter under review by the panel member.
q. Involvement with document under review—a brief description of any previous involvement of the panel member with the development of the document (or review materials) the individual has been asked to review.
r. Other potentially relevant information—a brief description of any other information that might reasonably raise a question about actual or potential personal conflict of interest or bias.

Further, the contractor shall require that panel members sign a statement that says the panel member is not currently arranging new professional relationships with, or obtaining new financial holdings in, an entity, which is not yet reported and which could be viewed as related to the topic under discussion or stakeholders associated with the topic.

Conflict of Interest Memorandum: Certification

U.S. Environmental Protection Agency
Conflict of Interest Inquiry

You have been requested by EPA to serve as a Peer Reviewer for _____, and your involvement in certain activities could pose a conflict of interest or create the appearance of a loss of impartiality in your review. Although your involvement in these activities is not necessarily grounds for exclusion from the peer review, affiliations or activities that could potentially lead to conflicts of interest are included in the table.

Please complete the table and sign the certification below. If you have any questions, contact [point of contact at EPA Office] at your earliest convenience to discuss any potential conflict of interest issues.

Conflict of Interest Analysis	YES	NO
a. To the best of your knowledge and belief, is there any connection between the subject topic and any of your and/or your spouse's compensated or uncompensated employment, including government service, during the past 24 months?		
b. To the best of your knowledge and belief, is there any connection between the subject topic and any of your and/or your spouse's research support and project funding, including from any government source, during the past 24 months?		
c. To the best of your knowledge and belief, is there any connection between the subject topic and any consulting by you and/or your spouse, during the past 24 months?		
d. To the best of your knowledge and belief, is there any connection between the subject topic and any expert witness activity by you and/or your spouse, during the past 24 months?		
e. To the best of your knowledge and belief, have you, your spouse, or dependent child, held in the past 24 months, any financial holdings (excluding well-diversified mutual funds and holdings, with a value less than $15,000) with any connection to the subject topic?		
f. Have you made any public statements or taken positions on or closely related to the subject topic under review?		
g. Have you had previous involvement with the development of the document (or review materials) you have been asked to review?		
h. To the best of your knowledge and belief, is there any other information that might reasonably raise a question about an actual or potential personal conflict of interest or bias?		
i. To the best of your knowledge and belief, is there any financial benefit that might be gained by you or your spouse as a result of the outcome of this review?		

CERTIFICATION

I hereby certify that I have read the above statements and, to the best of my knowledge and belief, no conflict of interest exists that may diminish my capacity to provide an impartial, technically sound, objective review of the subject matter or otherwise result in a biased opinion.

_____
(Name – please print)

_____
(Signature)

_____
(Date)

What Did You Think?

We strive to constantly provide the highest level of value for you. Please take a few minutes to tell us about your experience using this product.

To be taken to a short consumer satisfaction survey, please click here or copy and paste the following URL into your browser:

https://www.surveymonkey.com/r/OSAconsumerfdbck?
product=Science and Technology Policy Council Peer Review Handbook 4th Edition
 October 2015

Thank you for your feedback.

Sincerely,

Office of the Science Advisor
United States Environmental Protection Agency
www.epa.gov/OSA@epa.gov

www.ingramcontent.com/pod-product-compliance
Lightning Source LLC
Chambersburg PA
CBHW081257170526
45165CB00011B/3323